TINKER GO WALKING!

Memories from a nomadic life

Anthony Dalton

a.k.a.

Tinker!

Anthony Dalton

Tinker the younger

TINKER GO WALKING!

TINKER GO WALKING!

Memories from a nomadic life

Copyright © Anthony Dalton 2019

All rights reserved. No part of this publication may be reproduced, stored in a retrieval system, or transmitted in any form or by any means. Electronic, mechanical, photocopying, recording, or otherwise (except for brief passages for the purposes of review) without the prior permission of the author.

ISBN: 9781076465900

Cover design by Steve Crowhurst

Sketches on cover and inside © Steve Crowhurst

Sketch on page 77 courtesy of Shutterstock.com

Photograph on page 81 courtesy Shutterstock.com

Photograph of First Cataract of the Nile, page 89, © Library of Congress

Author photograph © Steve Crowhurst

Maps by Steve Crowhurst

All other photographic images © Anthony Dalton

Published by Anthony Dalton in conjunction with Amazon/KDP

For

My intellectual muse
With love and appreciation

Having been condemned by nature and fortune to an active and restless life...

Jonathan Swift – *Gulliver's Travels.*

... I became a Nomad: a person of no fixed abode. A wanderer. Or, in my case, a walker in search of adventure and knowledge.

Where's Tinker?
He's gone.
Gone where?
Oh, gone walking, I suppose. Gone to see the world. Timbuktu, or someplace like that.
Will he come back again?
Yes, I expect so, one day.

Yeah! Tinker go walking!

CONTENTS

PART ONE	BRITAIN	1
PART TWO	MIDDLE EAST	17
MAP: MIDDLE EAST		18
PART THREE	AFRICA	61
MAP: AFRICA		62
EPILOGUE		155
ABOUT THE AUTHOR		159
MORE BOOKS BY ANTHONY DALTON		161
PRAISE FOR ANTHONY DALTON'S WORK		171

TINKER GO WALKING!

AUTHOR'S NOTE

At a writers' festival in British Columbia an interviewer asked me why, after all my adventures, I hadn't written my life story. I had been asked this question before, a number of times, usually by my readers, without taking it seriously. I had committed parts of my life to paper, in some ways. I had written three autobiographical books, although two were about specific events in my life and the third was a collection of tales from my travels. One summer a few years ago, for the first time, I started considering the possibility of writing an autobiography, or a partial autobiography, with something akin to professional interest. As the memories piled up and my thoughts on the potential project became more structured, I realized I was already writing the story in my mind. The question here is quite obvious: is this Tinker's story or is it mine? Is Tinker, in fact, me in disguise?

I have been told some of my adventures and exploits are unbelievable. The truth, however, is often stranger than any fiction I could write. Therefore, I leave it to my readers to decide for themselves whether my stories are fact or fiction, or whether they are a mix of both. Whatever you decide, I hope you will enjoy travelling these ancient trails with Tinker and I on our personal odyssey through many exotic lands and a variety of historic time zones.

The adventures in this book took place in 1959 and 1960 when the world was far different from today's version. International boundaries were different. Place names were different. Mass tourism was in the future, as were cell phones and the internet. The world was simpler and, for some of us, safer and more interesting.

TINKER GO WALKING!

TINKER GO WALKING!

TINKER GO WALKING!

TINKER GO WALKING!

TINKER GO WALKING!

TINKER GO WALKING!

PART I
BRITAIN

TINKER GO WALKING!

CHAPTER ONE

Six Boy Scouts, all in uniform – black socks with red garters, navy blue shorts, navy blue short-sleeve shirts, red neckerchiefs bordered in white, tan broad-brimmed hats and, instead of the regulation black shoes – hiking boots on their feet – all in their early teens. Each one carries a long wooden staff, a notebook and a pen or pencil. They are on a weekend hike in the countryside of southern England. Their objective is to complete a circular route of almost ten miles while observing as much as possible of life around them. I am in that troop, but not the leader.

At first, we chatter in excitement. Soon we quieten down and begin to notice nature: a sparrow's nest in a hedge – three tiny eggs, a red fox loping across a field, a small dead snake on the side of the road, the first buds of spring on a willow tree, a fish jumping in a pond, a clump of wild hollyhocks in bloom.

Looking back, from the vantage point of experience, the hike was a clever way for our scoutmaster to teach us observational skills we would need throughout our lives. Many years later I remembered that day as a long walk during which I separated myself from my fellow scouts by taking longer strides. The proximity of others made the experience less acute for me. I needed my own space. I was happy out in front alone. Not as the leader: being in charge was not important, being alone was. I liked being alone. I was good at being alone. I liked being alone on foot in the English countryside.

Yeah! Tinker go walking!

Walking was much better than being at home under the family roof where we, the three kids – me, plus a younger brother and an older sister, had to be constantly alert for our mother's frightening mood swings. There was a much younger brother, but he was too close to babyhood to understand.

Mothers are supposed to be warm and loving and kind, especially to their own offspring. My mother wasn't, although she believed herself to be a paragon of the matriarchal species. In reality she was a tyrant who spoke with clenched fists and a viper's tongue. He, the man who spawned me (I hesitate to introduce him as my father because he failed miserably in that role), was weak-willed, an indecisive wimp in modern parlance, forever under mother's thumb.

Small wonder then that I left home at the tender age of almost three years. That's when my nomadic instincts first kicked in, I suppose. I wandered out of the Scottish garden in front of the house we lived in near Broughty Ferry and off into the countryside. How many adventures did I enjoy that day of freedom? I have no idea. I'll bet I had fun. I don't remember the occasion, but I have been assured it really happened.

I was found some hours after I went missing, after a frantic search through hedges and over fields, in a village over one mile from my home. I must have taken in excess of 5,000 little paces to get there.

"Tinker go walking," I explained with pride when I was found.

Recalling my mother's propensity for violent reaction, I imagine she slapped my bare legs in retaliation for worrying her. That was her way. She saw nothing wrong in that.

I've been told I was born at 08:20 at the tail end of an air raid a few miles east of London docks on the south bank of the Thames River during WWII. The midwife wrapped me in a soft blanket, thrust me into my mother's arms and then she fled to the safety of an air raid shelter dug into the back garden where others were already cowering from the bombs. The midwife is reported to have said something like, "I left the baby with her. If the house gets hit they might as well go together."

The house didn't get bombed but many others did that morning. The house across the road disintegrated from a direct hit. Amongst the rubble strewn across the road was a significant part of Mr. Johnston's left leg, and sundry other parts. He was an old widower who had lived in that house all his life and died in it when the bomb hit. We never suffered more damage than broken windows. In that we were more fortunate than so many others.

Ignoring the government's warnings to restrict domestic travel – "Is your journey really necessary?" the placards asked – we were a nomadic family. We moved from Kent to Norfolk and from Norfolk to Dundee in Scotland. We moved again, living for brief periods in Stirling, Perth and then to Broughty Ferry, close to Dundee. Later, in the last year of the war, we lived with another family in Dorset in the south of England. Always, it seemed, we lived on the charity of others.

Despite the odds against us through the bombardment by Germany's aerial might, we were among those who survived the war. We were the lucky ones, left alive to watch as Europe embarked on a period of uneasy peace.

The war was over. A time of readjustment. A time of peace. A time of tranquility. A time of continued frugality. Some of that is true. There were no more bombs dropping on us. No fighter planes from opposing nations battling in the skies above us. No more deadly doodlebugs, as the V2 rockets were known. So, yes, it was rather tranquil, I suppose, though not as peaceful as it could have been.

I was five when the war ended. War was all I had known. Noise and fear. Bombs. Wrecked houses. Not an auspicious start to a few decades on Earth. I spent the next six years getting used to a different life, a life without that kind of fear, as the second war to end all wars faded into recent history. And there it would stay, until the next time.

My uncle called me Tinker because he noticed, from an early age, I had to know how things worked. That was even before my Scottish solo adventure. Some labelled me a destructive child to see me surrounded by parts of toy cars, or by a dismembered doll from my sister's collection. He, my uncle, understood and watched as I carefully reassembled each vehicle with its correct chassis, engine and wheels, or each doll.

No one ever called me by my real name in those days, or for many years after. Uncle used to pat me on the shoulder and tell people, "He's a clever kid, this Tinker."

Tinker I was then, and Tinker I have stayed, to a select few. I'm a tinker. Tinkering is what I do.

At eleven-years-old I won a scholarship to a grammar school. No one in my family of aunts, uncles, cousins, brothers or sister had ever achieved so much. The tuition was paid for. A school uniform and all the extras, such as clothing for sports, were not. That ended the grammar school possibility as far as my parents were concerned.

"Well, he can't go. He'll just have to go to Temple Cowley School like everyone else."

I didn't have to do anything of the kind. Thanks to Mr. Halsey, a teacher from my junior school: a teacher with a heart of gold, I was guided to apply for a grant and I got it. The problem of extra expenses for grammar school faded. Problems at home continued.

My parents failed to understand the concept of homework. I went to school to learn. I didn't need to do it at home, was their stand. I was expected to help look after my youngest brother in the evenings and on weekends. I often did my homework before delivering newspapers in the early mornings when the house was quiet. My parents never asked what I was doing so early. I doubt that they could have helped me study anyway. I already knew more about many of the subjects than they did.

I was proud of the new information being filed in my mind. My pride wanted me to share those facts with my family. That proved to be a mistake. I was too young to comprehend the fragility of an underdeveloped adult mind. I interrupted my parents' conversation one day by contradicting one of them with, "Actually, that's not true…"

The reaction was startling. It was cruel. It became painful.

"What did you say?" snarled mother.

I repeated my comment and started to explain where they were wrong. I didn't get very far.

"You toffee-nosed little sod!" spat from father's mouth.

"Who the bloody hell do you think you are?" Mother's contribution followed by a hefty clout on the side of the head. "You think you're better than us!" The last shouted with quivering fury.

I wanted to scream out through my sudden tears, "No. You think I'm better than you. That's why you're so unpleasant. You're jealous of me and my knowledge."

I didn't, not then. That bravery would come later. Rather than risk an escalating confrontation, I rubbed the side of my stinging head, and fled the house for the safety of the outdoors. So much for the wonders of education, I thought. Dreams were so much safer.

My dreams gave me the strength to accept what I couldn't change and to undertake difficult tasks as a boy, and later as a man, and to understand when I failed in some of my quests.

My days started at five in the morning. I woke up, washed, dressed, did any homework necessary, made myself a cup of tea, toasted two pieces of bread, buttered them to make a warm sandwich, got on my bicycle and rode to the local newsagent – about a mile away, while eating breakfast. There I sorted and counted a variety of newspapers, from the cheap and argumentative *Daily Mirror* to the classy erudition of *The Times* and *The Guardian*. Leaving my bike at the shop, I walked my route. In those days that meant opening garden gates, walking the length of the garden, pushing the paper or papers through the letterbox, walk back to the gate, close it quietly, and on to the next house. The routine never varied. Rain, shine, sleet or snow. I walked the same route with the same newspapers seven mornings each week. When I had finished, I raced home. On weekdays I stripped off my clothes, put on my school uniform and cycled to my studies.

On Sundays I was a choirboy; had been for years. A member of the full choir and an altar boy for the morning service, I became a soloist for Evensong.

My combination of school and working life didn't end with studying and delivering newspapers. After school I worked for a pharmacy from Monday to Friday, delivering prescriptions by bicycle for one hour. Only then could I go home. The money I earned from both jobs went into the family fund. I kept no more than about twenty percent of my pay. That money, however, gave me great pleasure. It allowed me funds to tinker with my increasingly skillful hands.

I designed and built model aeroplanes from balsa wood with basic tools and I taught them to fly. I fabricated my first guitar from scrap plywood (it sounded awful). I built small cars and trucks out of odd pieces of wood for my youngest brother. I also repaired bicycles for other kids – straightening buckled wheels, mending punctures, tightening chains, adjusting saddles. I enjoyed tinkering.

I met a real tinker once. He was a gypsy with an accent straight from the hills of Connemara. He and his wife, I assumed that's who she was, travelled the land in a colourful wagon with a rounded top (they called it their caravan) pulled by a powerful cart horse. I saw them trundling along a country lane as I strolled home one weekend from fishing in an old quarry near Dorchester. I waited until the horse drew alongside me and then I fell into step, although I walked backwards to look at the gypsy and his wife.

"Hello," I greeted them with a smile. "Who are you and where are you going?"

"'Tis a tinker oi am," he replied with a grin. "I repair pots and pans and anyting else made of metal."

"Anything?" I asked.

"Anyting," he answered.

"Where are you going?"

He pointed beyond his horse's ears. "That way," he said.

"And where have you come from?"

He jerked the thumb of his left hand over his shoulder and sang out, "Back yonder, me boyo."

I laughed with delight. That was my kind of answer. He was my kind of character. I wanted to climb up on the wagon and travel with them. He grinned at me and asked, "What's your name, young fella?"

I hooted with laughter again and told him, "It's Tinker. Tinker Taylor. I'm a tinker too, just like you."

"Go on with yer blarney," the tinker's wife called, a smile lighting up her face; her brown eyes asparkle. She was a dark-haired, dark-eyed beauty in a flowing skirt of black silk decorated with vibrant reds and greens. She wore a ruby red top partly covered by a short black jacket with long sleeves. Her hair was tied up in a bun wrapped in a red and green scarf that trailed over her shoulder like an invitation to the future. She was stunning and she knew it. I was only twelve years old, but I fell in love with her right there and then. I think she knew that too. Love at first sight.

I grinned up at her. I couldn't have concealed my thoughts, even if I had tried. She blew my smile a kiss, clucked her tongue against her teeth, took the reins from her old man's hands and shook them. The horse was of a different mind. It shook its head. "No chance," it seemed to say, and continued at the same slow gait.

"If you're going that way, you must be going to Salisbury," I pointed over my shoulder to the north-east. "I don't think you can take the horse and caravan through the city. There's too many cars and bicycles."

The tinker grinned at me. "If Salisbury is that way then that's where we are going. Perhaps we'll be after learning something there."

"But, what about the traffic? The horse won't like that."

"Ah, to be sure, the *aos sí* will take care of us." He pronounced the unfamiliar words as ees shee.

"What's that? The – what?" I tried the pronunciation and slurred my tongue. "What's the, what you said?"

"The *aos sí* are the faeries. They will see we come to no harm."

"Faeries?" I tripped on a stone and fell on my rear. The tinker and his wife laughed. He gave me a thumbs up sign as the wagon creaked past. She blew me another kiss. It felt like faery dust.

"You believe in faeries?" I asked. I got up and ran a few paces to catch them, my mind filling with exotic images of little people dressed as butterflies."

"Aye, we do, to be sure. *Go n-éirí leat*," he called. I didn't understand until many years later that he had just wished me good luck in Gaelic.

Happy with the encounter, I watched them until they were out of sight. A tinker's life, I thought. That could be me behind that horse. That could be me out on the open road living in a coloured wagon with a beautiful lady and looked after by faeries. Ah, a tinker's life for sure.

Instead of going a-tinkering, I went home. I opened my creel, placed two pan-sized rainbow coloured perch on the kitchen counter and told mother, "You might like those fried, when Dad gets home from work."

Mother sighed. She did that a lot. All the worries of the world on her shoulders. I hated that theatrical nonsense. "And who's going to gut them and clean them?" she asked. "Not you, I'm sure."

"No, not me. I'm not allowed to use sharp knives. You said so. Remember?" I gave her a cheeky smile from a safe distance. "I'm off. I have homework to do," I continued as I backed out the door.

Her shout of annoyance followed me up the stairs to the room I shared with my younger brothers. They were playing snakes and ladders on the floor. They were being noisy.

"Be quiet, little people," I ordered as I walked in. "Big brother has work to do."

The youngest ran out calling to his mother that I was being mean. The other glared at me.

"Prick," was all he said.

"Watch your language," I cautioned.

"Prick," he said again, adding a glare for emphasis. I knew he was trying to provoke a fight that he could not win. That never stopped him trying though.

I ignored him as I dropped on my narrow bed and opened *Gulliver's Travels*, my favourite book at the time. I had an essay to write for school on an imagined journey to a land of my choice. I chose fantasy. Jonathan Swift was my guide. He, like the tinker, was Irish. He, like the tinker, knew about the little people. I let him guide me to the fabled land of Lilliput, cast adrift in the Indian Ocean. I would base my essay on my version of life in a land like Lilliput – without the offbeat laws.

One of the silliest dictates in Gulliver's Lilliput was the correct way to open an egg before eating. The very idea was hilarious. Due to a simple accident to a relative, the emperor commanded that eggs should only be opened at the smaller end, upon pain of death. Six rebellions resulted from that edict and, Lemuel Gulliver reported, 11,000 people died as a result. All because an egg was cut at the wrong end!

My youngest brother, many years my junior, claimed he didn't like the top of a boiled egg. It didn't taste nice, he insisted, so mother turned his egg upside down, cut off the rather pointed smaller section and he accepted it without complaint. I shook my head in amusement. Making sure he was watching, I studied both ends of my boiled egg, placed it in the eggcup with the pointed bottom down and, having no fear of the Emperor of Lilliput's wrath, sliced off the top of the broadest part. I then ate it with much smacking of lips. Younger brother immediately began to cry.

"I don't like my egg like this," he sobbed. "I want one like Tinker's."

Mother swore at me. Too far away across the table for violence, she yelled, "Wait till I get hold of you, you little sod."

I didn't wait. I took my egg and spoon and I ran for the sanctity of an open field to dine in peace, and to look for little people, perhaps wearing butterfly wings, who might appreciate my attempts at satire.

Compared to others of my age, I was a quiet boy. I was a shy boy. I was a loner. I was uncertain and often nervous – probably a hangover from the war years and my mother's black moods, but I had dreams. Those dreams gave me the courage to talk to the tinker. They gave me the courage to seek out the faeries. They gave me the courage to begin to believe in my own capabilities. I was happy with my own company.

A stone railway bridge, carrying a single line on its back, stood over a narrow lane not far from home. I enjoyed watching the steam trains, wondering where they were going, wishing I could go with them. The bridge, though, was a static challenge. The walls had a slight incline due to the base being more substantial than the top. I hoped to join the gymnastic team at school one day. That bridge gave me the opportunity to learn physical control. That bridge taught me to climb. I practiced whenever I had free time on weekends, learning to use my fingertips and the toes of my running shoes to inch from stone to stone; from crack to crack. The bridge was no more than thirty feet high – enough for gravity to seriously injure me if I made a mistake. I never did. I couldn't afford to.

I spent many happy afternoons on an abandoned WWII airfield. There I cycled alone for hours. I walked alone for miles. I watched butterflies at close quarters to see if they were insects, or faeries in disguise. I was almost convinced that the Monarch was a faery. It had to be. Monarchs had deep orange wings edged in dark brown, or black, with a scattering of white polka dots on them. I was sure they had to be faeries. I didn't share my knowledge with anyone: boys or girls. My thoughts were my own. I never did find any faeries, so they receded to the back of my mind, but they have never truly left me.

I was surprised at that time to learn that many girls like quiet boys – studious loners such as a boy called Tinker. Perhaps they have an instinct to mother them. As I meandered through that first of my teenage years, I began to look at girls the way I had looked at the gypsy lady. Changes were going on inside my body and in my mind. The next stage of my education began with a firework display in my head and electricity in my loins when an 18-year-old encouraged me to search for her in a thicket beside a golf course. I found her easily. I think I knew she wanted me to. She, Molly, initiated me into the differences between girls and boys.

A year or two before I had asked mother why boys and girls were different. I didn't know for sure that they were, but I had heard rumours. I wanted more information. Not just, were they different? Or, why were they different? But how they were different? Mother didn't believe in talking about such things. Smutty talk, she called it. She avoided the topic with predictable annoyance. "Oh, ask your father. I haven't got time."

I knew that wouldn't work and I said so. "He won't know. He doesn't know anything. He just puts one finger against his lips and says, 'Well, I'm not sure.'"

True though my observation was, that earned me a slap on my ear. I went looking for answers elsewhere. That's where Molly eventually came into my life. She was taking a short cut home from work across the golf course on her bicycle when I intercepted her. I was up in the fork of an old oak tree watching a barn owl watching me. Molly cycled towards me. I swung upside down, doing my Tarzan impression, hanging by my legs, and stopped her.

"What the hell are you doing, young Tinker Taylor?" she cried. "You scared me."

She looked different upside down. I remember thinking that, if she were upside down, her skirt would be hanging the other way and her legs would be bare and I would be able to see her underwear. The thought gave me a funny feeling inside. I forgot about the owl, turned right side up, let go of the tree and dropped to the ground in front of her. She stood beside her bike. Her skirt hanging the way it should.

"Nice bike," I said, (gauche, I know, but I was just a kid). I was looking at her body, trying to imagine it without clothes, not looking at the bike, or into her eyes. Without clothes, how different would her body be from mine? Would it be different? I wanted to know. Molly saw what I was thinking. She started to laugh and propped her bike against the tree.

"Do you want something?" she asked. "Where are your friends?"

"There's no one here," I answered. "I'm by myself. I'm always by myself. I like it that way." I leaned back against the bole of the tree and told Molly, "You look nice."

Molly smiled and looked around her. "Thank you. Do you want to have an adventure?" she asked. Of course I wanted adventures, I thought. That was my language.

"Okay," I said.

"Close your eyes and I'll hide. Then you have to find me," she ordered

I took my time. I knew where she was hiding. I had heard her rustling around in the autumn leaves. I found her on her back, her loose skirt above her knees, which were raised, showing her long, smooth, bare legs. She patted the dead leaves beside her.

"Come here," she invited. Less than an hour later I knew all I needed to know about the anatomical differences between us. I learned a lot more, too. I learned how to kiss properly for a start. That was fun. A bit dribbly at first until I got used to it. We moved on from there as Molly guided my hands over the exciting, soft, smooth new terrain that was her body.

"Touch me here," she suggested. "Kiss me here." I was learning.

She met me about once a week for a few months after that. I maintained her bicycle for her. She educated me. I always had to brush leaves and grass out of her auburn hair and off her back before she went home. My dirty elbows and knees were my boyhood badges to denote my outdoor pursuits. No one thought anything unusual about them. No one, except Molly, had any idea that I had passed through an important portal in my life.

Molly reminded me of the gypsy lady. Not because of her looks, or the clothes she wore. No similarity there. To be honest, I can't even recall what Molly looked like. But the gypsy lady's face and style stayed with me. No, it was more than that. After that first exciting experience on the edge of the golf course, I knew that the gypsy lady must be just like Molly under her flamboyant clothes, although, I suspected, somewhat more exotic. It was my fantasy. I could picture her however I wanted. I hoped one day I would find a gypsy lady of my own. Until then I had to survive home life.

In my mid-teens I sat with my youngest brother on a weekend telling him tales from a book of adventure. Mother overheard and shouted, "What rubbish are you filling that child's head with now?"

"It's not rubbish," I answered, not looking up and so missing the warning signs. I held up the book to show the front cover. "It's a Greek classic. A travel book. It's beautiful…"

That's as far as I got. My much-read, much-loved copy of Homer's *The Odyssey* was ripped from my hands and thrown across the room.

"That's my book," I said, keeping my voice as controlled as possible under the circumstances. Mother's face grew red. She clenched her fists. "You. You… You…," she stuttered in her anger. Here it comes, I thought. "You think you're so much better than the rest of us."

There it was again. The underlying cause of much of the friction in the house. My education was the problem. This time I said what I had only thought on so many other occasions.

"No, Mother. The problem is, *you* think I'm better than the rest of you and *you* don't like it."

I moved my kid brother aside and stood to face the blow I knew had to come. The clenched fist wavered, shot out aimed at my head. I caught her wrist in one hand and held her away from me. She struggled and spat out, "You think you're a man now? You're not a man. Real men don't hit women."

The sad part of that scene is that she actually believed she was in the right.

"I didn't hit you, Mother. I stopped you from hitting me."

"Don't argue with me," she hissed – the viper's tongue. "Let me go."

I let go of her wrist. Before I could duck she had punched me in the face.

"There!" she shouted in triumph as I staggered back. "There!"

I licked my lip and tasted blood. Noticing that her blow had cut me, she put one hand to her mouth and said, "Oh, I didn't mean that to happen." No apology, just a weak explanation.

As always on such occasions, I retreated to the outdoors to be alone – there to heal the wounds and hide the scars. The fields and forests, the country lanes and the riverbanks were my sanctuaries.

I joined the air cadets at 15. That gave me one evening a week to learn more mechanics – another sanctuary. School, too, became a hideaway for me.

At school I earned a place in the gymnastics team and I gained a place in the cross-country team. In the classrooms I tried hard not to daydream, always a failing for me. In history we were studying the impact of Greek culture on the world, and we were studying the British-led industrial revolution of the 17th and 18th centuries. The industry lessons satisfied me to a certain extent, but the development of factories, steam trains, and the inevitable agrarian reforms could not compare – in my mind – with the romance of the wonders of antiquity. Even though I was a tinker with a bent for mechanics, ancient Greece, the Persian Empires, Egyptian Pharaohs, Imperial Rome, they were the subjects that inspired me. I determined that, as soon as I had enough money, I would leave home and go to the sources to learn more.

The family emigrated to Canada when I was seventeen. Further away from my goals for me, but an exciting move, nonetheless. The upheaval, a new land, the strangeness, affected mother more than any of us could have imagined. She was homesick for England within minutes of arriving in Toronto. Family life did not improve.

I left home again in my late teens – a year after moving to Canada and six years after Molly had worked her magic on me. Home wasn't a happy place. There was no reason to stay.

I said I was going to Timbuktu, and I was, but I was also on a quest for knowledge. That quest would take me far to the east studying a young royal Macedonian warrior. It would take me south to where architecture and intelligent thought had defined a great civilization beside a venerable river. It would take me across a vast desert to a mythical town where ancient intellectuals studied religion in a university older than Oxford or Cambridge.

On the way, inspired by Mr. Swift's elegant prose and exciting mind, I went looking for my own versions of Lilliput and Brobdingnag. My personal Lilliput. My personal Brobdingnag. Nothing to do with the physical size of the people, more a desire to learn their cultures. No one came looking for me that time. I was free, as free as the tinker and the gypsy lady I had once almost known.

"Where's Tinker?"

"He's gone."

"Gone where?"

"Oh, gone walking, I suppose. Gone to see the world. Timbuktu, or someplace like that, he said."

"Will he come back again?"

"Yes, I expect so, one day."

A hesitant answer steeped in uncertainty. No one knew the real answer. Not even me.

TINKER GO WALKING!

PART 2

THE MIDDLE EAST

TINKER GO WALKING!

TINKER IN THE MIDDLE EAST

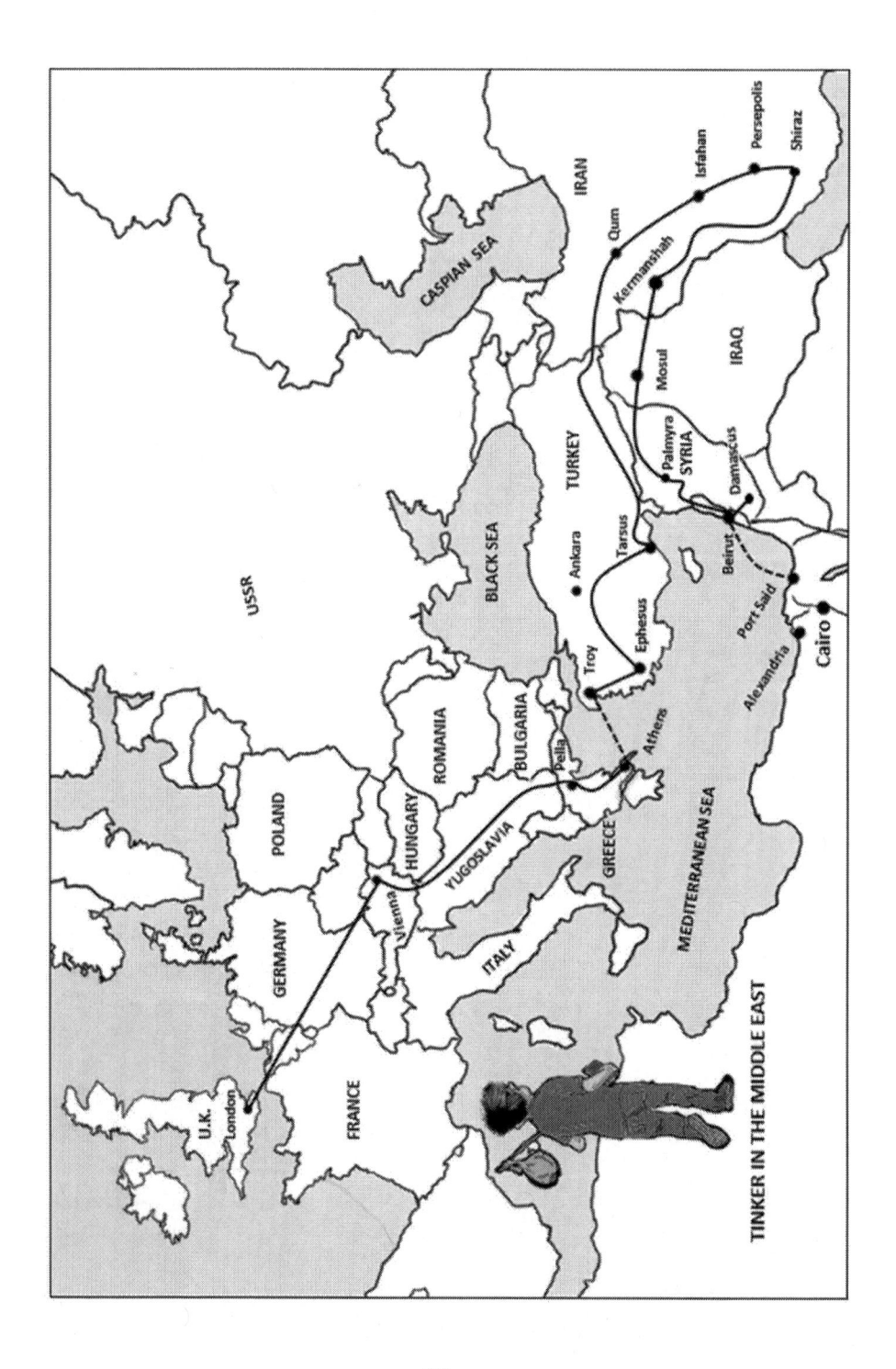

CHAPTER TWO

Alexander the Great was one of my early heroes. I wanted to know more about his life and his travels. That was part of the impetus for my own journey of discovery. He had nothing to do with Timbuktu, of course. Alexander campaigned across the earth centuries before the legend of Timbuktu grew out of the Saharan sands. Even so, there was a tenuous connection between the two. I was that link. My ultimate goal was to reach Timbuktu: on the way I wanted to learn as much about Alexander as possible – that entailed a massive detour through ancient Persia, walking in his footsteps where possible. When I set off I had no clear knowledge of the other historical characters I would collect along my path. Discovering them was a bonus. Learning of the links between them was educational and it was fun.

Cleopatra was one of those additional characters. I've often thought about Cleopatra. I've seen ancient busts that supposedly represent her beauty. I've seen her likeness on coins and friezes. The Vivian Leigh representation in an old movie was stunning, no doubt about that, even so, none of those effigies could truly prepare me to meet Cleopatra face to face, or with any spark of recognition. I solved that dilemma by sharing a gypsy lady's face with her. That made her real to me. That made Cleopatra accessible. The gypsy lady from my childhood had become my icon. I took her everywhere with me in my head and employed her as I wished. She would be my Aphrodite and my Diana. She could be my Cleopatra as well. My mind was happy with that subtle deception.

I'm guilty of mentally misrepresenting Cleopatra, as, I suspect, are many other people. I thought the powerful lady who ruled Egypt and became a mistress to both Julius Caesar and Marc Antony was an Egyptian, daughter of a long line of illustrious leaders from the fertile Nile Valley. Not so, I learned. Cleopatra's antecedents were Greek Macedonians. Her father was Ptolemy VII and her mother was Cleopatra V. Our Cleopatra spoke Greek fluently as well as Egyptian. Cleopatra was closer in family ties to Alexander the Great than to the early pharaohs of the native Egyptians.

I added Alexander to my study of Cleopatra. Antiquity had taken on an intriguing new dimension. How many Cleopatras had there been? I asked myself. I had no idea. A visit to a library with volumes in English would supply the answer. That meant Athens and the excellent National Library of Greece.

A few weeks before I had flown from Toronto to London in early September and leap-frogged from city to city making my way across Europe 'on the thumb' as the expression went in those days, my focus entirely on reaching Greece, I often slept rough. Occasionally I stayed in Youth Hostels. I fed my grumbling belly by helping to repair flat tires, start motorcycles, cars, tractors, trucks, buses, fix bicycles, and sundry household appliances. Tinker, tinkering his way to knowledge.

Soon after leaving Vienna, where I had eaten a sausage from a street vendor near the railway station, I had a violent attack of the trots. It lasted three days during which I slept on hay in a rat-infested barn near Zagreb and lost a lot of weight before I moved on.

The wonders of ancient Greece kept me enthralled longer than I expected. I went looking for Alexander at Pella, in Greek Macedonia, soon after I crossed the border from Jugoslavia, and found Aphrodite as well. She was beautiful in all her poses, but she looked so sad. Why, I wondered, did the sculptors of long ago form one so beautiful to be so mournful? Think about it. Aphrodite was the epitome of sensuality. She was sexy. She was lovely. She was wise. She was a goddess. Surely she must have been happy, at least some of the time. Surely she knew how to smile? I sat in wonder, studying the beauty of her life-size form. Her body was perfect. Her curves so smooth. I stood and approached her. She didn't move. No hint of a smile on that troubled face. I ran my left hand over her right hip, feeling the warmth from the sun on her splendid body. On a whim, I leaned forward and kissed her on the lips. Was there a sudden flicker in her eyes? Did her lips move a fraction in response to mine? Logic says, no, that could not happen. But I was there. I know what I felt. Aphrodite had recognized me.

Aphrodite is said to have been the wife of Hephaestus, God of blacksmiths and metalworkers. A tinker, in other words. Hephaestus was a tinker just like me, except that he was a Greek God and he was lame. Aphrodite, Goddess of love and beauty, was not only a wife, she was also a philanderer. Among her many extra-marital lovers were the handsome but unpopular Ares – God of war, and Anchises – a Trojan shepherd of diluted royal blood. I didn't really care how many lovers she had taken to her bed, in or out of marriage, she was a beautiful lady with a sensual figure. I was a testosterone laden young man. I could not feel sympathy for cuckolded Hephaestus. Aphrodite was my kind of woman.

Ancient Greece fascinated me. Had it been possible, I could have stayed in the Greece of a long-ago time and spent years listening and learning from the great philosophers. Modern Greece had its appeal, but the politics bored me. History and understanding were my reasons to be there. Aristotle, Socrates and Plato lived as intellectual shadows on the paths I trod. Aristotle, Socrates and Plato, and, of course, Aphrodite, who summoned me to follow her. They were my scholarly and sensual guides to their ancient land of rational thought. I roamed south from Pella to Athens to study in the library until my mind became sated.

In Athens, after days of happy hours in a quiet reading room, I found the Ptolemy clan. Among them were Cleopatras I, II, III, IV, V, VI and VII. A redoubtable list of strong, successful women. How could I roam the centuries in my short lifetime and know them all? I couldn't, so I compromised and settled on the most famous one. The one who looked a lot like Vivian Leigh. Cleopatra VII, the one who seduced two powerful Romans – Julius Caesar and Marc Antony. Cleopatra VII, the last ruler of Egypt from the Ptolemy dynasty before the might of Rome took charge. I carried Cleopatra with me in my mind.

My Athenian days passed mostly in the library. I spent one complete day outside on the Acropolis, exploring the Parthenon and imagining the erudite conversations that must have taken place on that great lofty rock. My evenings were reserved for cheap, tasty meals, red wine and traditional Greek music in the Plaka.

As the young Alexander learned from Aristotle, so I learned philosophy from them both. Aristotle told Alexander, "It is the mark of an educated mind to be able to entertain a thought without accepting it." Alexander lived by that dictum and, then as now, the profundity of the statement became a mantra for an itinerant Tinker.

Socrates was confusing. He seemed to be little more than a literary figment of Plato's wisdom. Did Plato invent Socrates, or did he really exist as the father of western philosophy? I wish I knew. (I can hear classical scholars screaming in anger. I don't care, this is my story. I can relate it any way I wish.) Even though Socrates was supposed to be the teacher, I shall bypass him and concentrate for a moment or two on Plato, Aristotle's muse and, therefore, Alexander's eventual inspiration.

Plato was born in Athens in 427 B.C. the son of Ariston and Perictione. His family tree suggests a familial link with Poseidon, the God of the Sea – who himself claimed to be a brother of Zeus and Hades. Reading much of the history of ancient Greece gives the distinct impression that claiming kinship with the gods was standard procedure for those fortunate enough to have received formal education or to have been blessed with extraordinary beauty. Where did that place me? I wondered. I had the education, but I didn't see much more in me to suggest I had the makings of a god, and no one else had mentioned the possibility. Perhaps elevation to the supernatural, in the manner of deification, was and is only for beautiful young Greeks of either gender, and for old Greeks with above average intellects

Whether or not he was a god in real life, Plato had a remarkable mind. His epic dialogue *The Republic* remains today the most erudite philosophical discussion on justice in all its forms. And there we meet Socrates again. He is the dominant character in *The Republic*. The book is filled with images of Socrates and the brilliance of his arguments. He claimed, 'the just man is wise and good, and the unjust man is ignorant and bad.' He also maintained, 'injustice produces internal disharmony which prevents effective actions.'

On the first statement I find it hard to agree entirely. Portraying the unjust man as ignorant doesn't work for me, although I agree that he is bad. I do agree with the second statement. His third argument, that 'the just person lives a happier life than the unjust person' is as debatable now as it was when Plato put those words in Socrates's mouth. They were put there for the joint purposes of intellectual thought and discussion. Plato left me with much to ponder as I roamed his world. We learned the term *platonic* – as in relationships – from Plato, although it is unclear whether he employed the concept in his philosophies. I could never find any such reference.

Leaving Greece, I followed Aphrodite's trail across the Aegean Sea to the ruins of Troy, although I stopped on the island of Chios for a night and a final Greek meal and stayed for a month. Idyllic days of spear fishing for octopus to sell to a harbour-front restaurant owner, turned into evenings waiting on tables for the same restaurant owner under the Aegean night sky. The days passed. The evenings hurried by. The weeks fled. Christmas came and went. A new year dawned. I was happy on Chios. I had to leave – or stay forever. My forever was already committed. I had to go.

Troy is in Asia, not far across a narrow body of water from Chios. Aphrodite's scent was there; so was Alexander's trail. That notwithstanding, Troy was a disappointment. There was a bloody great wooden horse, of obvious recent manufacture, towering over the ruins. Austrian archaeologist Heinrich Schliemann would have been appalled, so would his much younger wife Sophia. They uncovered significant parts of Troy in 1870. Finding a huge wooden effigy of the Trojan Horse on site as a tourist attraction would not have given them any intellectual pleasure. The hollow horse didn't do anything for me either. I looked at the tourist trash in fascinated horror before turning my feet away towards the south. Disappointed beyond measure, I wandered towards Ephesus and my first physical appointment with a lady named Cleopatra.

Now historical time became confusing. I was on a trail marked by the feet of Alexander's considerable army which set out from Pella in 336 B.C. I was also looking for the most famous of the Cleopatras – Cleopatra the VII who ruled Egypt from 51 B.C. (on the death of Ptolemy, her father) until she committed suicide 21 years later. That 300-year differential between the lives of Alexander and Cleopatra should have caused me concern, but I took it all in my stride.

The white marble ruins of the architectural beauty that was once a thriving city named Ephesus sit on the side of a hill overlooking the Aegean Sea. Ephesus has watched over its part of the Aegean coast since before the 7th century B.C. From an insignificant Neolithic settlement, it survived through the Bronze Age, and the Ionian era. And then, in the 4th century B.C., General Lysimachus, one of Alexander's high-ranking army officers, established a new Ephesus close to the old site. The town grew and the centuries passed. A parade of Hellenistic princes ruled one after the other until Ephesus came under Roman influence in 133 B.C.

The Romans didn't last long. They were ousted 45 years later by King Mithridates of Pontus (a large territory in northern Turkey bordering the Black Sea). Mithridates didn't care for Romans or for their statuary. He massacred the Roman inhabitants of Ephesus and had all Roman statues destroyed. Mithridates managed to rule for a miserable two years before the Ephesians tossed him out and again embraced the Romans. General Lucius Sulla showed up a year later to take charge. He was a skilled administrator who organized overall Roman influence in Asia Minor.

There was an old sand-coloured British registered ex-army Bedford RL 4-wheel-drive truck parked near the entrance gate of Ephesus with both doors of the cab wide open and muffled curses emanating from the engine. A pair of booted feet waved at me from the passenger side. I stood clear of the boots to avoid being kicked and called out, "What's the problem?"

More muffled curses from the vicinity of the engine. The booted feet wriggled out the door followed by a pair of legs, shorts and a torso wearing a dirty shirt. A face, smeared with oil and topped by a mop of light brown curls, peered at me.

"Yes, I have a problem." He had a classy accent and he sounded pissed off. He was holding an oily fuel pump in one hand and an adjustable wrench in the other. "It appears my fuel pump has gone on strike."

I took the pump from him and looked it over. It was a repair job tailor-made for a tinker.

"This gasket has fractured," I said, pointing to the problem. "Fuel is leaking out here."

"Where will I find a gasket for a pump like this?" the owner moaned.

"Don't you have any spares?"

"No, not for this."

I noticed he was wearing a wide leather belt. It looked to be about the required thickness and width.

"A garage will charge you an arm and a leg to repair this pump and re-install it. I'll do it for two pounds, if I can have your belt as well."

He looked at me with a hurt expression. "Why do you want my belt?"

"I just want to cut three inches off the end – to make a new gasket," I explained. "You can keep the rest."

The penny dropped and he smiled. "Oh, oh, I see," he said.

An hour or so later his Bedford started with a distinctive diesel clatter. The pump was no longer leaking. I stashed two one-pound notes into my money belt and the owner drove away to an untold destination, with a foreshortened belt holding up his pants. His fuel pump would last for a few months longer and I could afford to live for another couple of weeks. We were both happy.

Having made someone else's day, and been rewarded for my efforts, I cleaned the oil from my hands, shouldered my pack and passed through the entrance for Ephesus. My first impression was that the engineers who laid the Marble Road could teach today's road builders much about permanence. The Marble Road, after a couple of thousand years of use, is in better condition than many city streets in Europe. There is even a road sign left over from antiquity. Carved into a marble paving stone, it is shaped as a footprint and believed to be a less than discreet sign to lead interested men to the brothel.

Paul the Apostle spent a couple of years in Ephesus (although it is unlikely that he took notice of the distinctive footprint mentioned above) a couple of decades after Jesus Christ had been executed in the land of Judea. Paul, converted from the non-believer – Saul, preached Christianity at every opportunity, perhaps even from the stage of the massive open-air theatre.

Cleopatra came to Ephesus before Paul, but I shall have to save that nugget for a little later, until I've shown you Tarsus.

The 2nd century A.D. proved to be a time of peace and prosperity for Ephesus and, indeed, for Asia Minor. An official Turkish guide, complete with an oval bronze badge on his lapel, told me the best parts of Ephesus were built during the 2nd century. He indicated all the identifiable buildings around us. He also told me no one was allowed to wander the ruins without an official guide. I knew that was not true and suggested, politely – I must add, that he should look elsewhere for a more innocent pair of ears.

Alexander, Aphrodite, and Cleopatra, collectively they were my focus and yet modern-day Turkey was there all around me to add its own flavour to my quest. I think of the salesman who drove me inland from Ephesus to Eskisehir and talked all the way. A young man, not much older than I, he chain-smoked cigarettes of doubtful manufacture. He looked at me instead of the road and he waved his arms in all directions to emphasize his monologue, which took place in a curious mixture of English, French and Turkish, and he somehow managed to keep his car going in the right direction. I had no idea what he was talking about. I do know he swept me through the countryside without my being aware of anything except the potential danger of his driving. We parted company on the outskirts of the town. He drove in for business appointments. I started walking – happy to be on foot again. Alexander marched his army along this route.

The Turkish hills leading to the Anatolian Plateau were cold in winter. I suffered snow, sleet and rain as I plodded up and down those hills. After walking many miles without being offered a ride, indeed, without even seeing or hearing a vehicle engine for hours, I passed through a village with the restful name of Abide (Ah- Bee-Day). I bought bread as the daylight waned and went in search of a comfortable site on which to sleep. A villager had warned me of wolves, maybe a bear or two in the hills. If they were there a campfire would keep them at bay – I hoped. I found a small grove of olive trees on a hillside overlooking the village. There I built a small fire, for warmth and for limited protection. Dinner was a meal of fresh bread and olives dipped in a little oil. I made tea in my billycan and settled down in my sleeping bag to watch a dusting of snow creating a thin veil over the land.

I drifted into dreamland and slept peacefully until I awoke with the sensation of not being alone. I could feel a warm pressure against my back. My mind made the obvious link from solitude to a bear – or a wolf. I turned over as slowly as possible, reaching for my sheath-knife. All around me were goats, asleep and awake. One greybeard showed interest, perhaps curiosity. His head bobbed up and down and from side to side, slowly. He wondered about me. I thought of him in terms of a spit over an open fire. The pressure against my back proved to be an old goatherd who had joined me for warmth.

I slid out of my travelling bed and pulled on my hiking boots. The goatherd sat up, rubbed his eyes, cleared his throat, spat into the fire, and grinned at me. "Merhaba," he said, poking at the embers to get the fire going again.

"Merhaba," I greeted him. I rubbed my hands and held them out to the fire for warmth. The snow had stopped. A thin layer of flakes spread over the hillside.

I had some bread left, and a few olives, plus the tea. He had a fist-size piece of feta wrapped in a dirty cloth. We shared our bounty while he talked to me in a soft voice about his goats and his life. I didn't understand a word of it, of course, but I watched his face, his eyes, his hands, and wove the story from there.

When we parted, as the sun changed the colours around us from grey and white to a more substantial olive-green, we shook hands, in the way of old friends. I stamped out the fire and poured the dregs of our tea over a glowing twig. He nodded in approval at my caution. Then, with a wave of his stick, and a cry to his goats, he headed up the hill to a higher pasture. I shouldered my pack and went looking for Alexander's spoor.

History forced me to a brief standstill in an unprepossessing community with the complicated modern Turkish name of Yassihuyuk. This, I knew, was the site of ancient Gordium. Here, in 333 B.C., Alexander encountered a chariot with its yoke tied to a post by a complicated knot with the ends hidden. Legend told that the knot could only be untied by the man who would rule Asia. Unable to undo the knot with his fingers, Alexander solved the problem with a warrior's gesture, rather than an intellectual appraisal. He drew his bronze double-edged sword and sliced through the knot. He went on to prove the legend by ruling much of the known Asia of that era.

Cutting the Gordian knot. Still in use over 2,000 years later as a metaphor to describe an easy solution to a complex problem. Clever.

I avoided Ankara, even though Alexander had passed through there. Modern Ankara would have been too cosmopolitan; too busy. Too noisy. I pointed my feet south, across the Anatolian Plateau towards the Mediterranean Sea. I rode on trucks and buses. I rode in cars, and for a hop from one village to the next, I lumbered along on a wheezing tractor with a non-communicative driver older than some of the antiquities I had seen.

In Konya I stopped long enough to bow my head at the ornate green- domed tomb of Mevlana Celal Ed-din al Rumi, an Afghan mystic from Balkh who founded the 'whirling dervishes.' The very name suggested something rather violent to me. A search for the truth acquainted me with something quite different. Whirling Dervishes were followers of the man from Balkh. They were monks who lived in cloisters. A Turk later explained to me, "The ritual whirling of the dervishes is an act of love and a drama of faith." Nothing too violent about that.

Today's Dervishes, adorned in long, wide, white robes to symbolize the death shroud – belted at the waist, and wearing hats of camel fur, spin like tops until they enter the desired state of trance, or they get dizzy and fall over – which must happen sometimes. The complete ritual is far more complicated than my basic description suggests. Needless to say, it is fascinating to watch although the dancers made my head spin with them. In retrospect, I think I prefer a sedate old-fashioned waltz.

The Turks are kind to travellers such as I. Rarely did I have to wait for long between rides on main roads. From the outskirts of Konya I had the comfort of a private car driven by a pleasant man who spoke not a word of English, or any other language apparently, apart from his native Turkish. For all that, he smiled, and he took me to the coast. I thanked him at Anamur and walked a short distance to Mamure Castle.

The original castle on this site is believed to have been built by the Romans in the 3rd or 4th century. Later owners, the Byzantines, the Crusaders and then the Seljuks added to and strengthened the imposing structure. It is recognized as the best-preserved medieval castle on the Mediterranean coast – and there are many of them.

Riding on top of a loaded truck gave me a superficial look at Silifke Kalesi, another castle close to the coast. It overlooks the Goksu River from a rocky promontory. Rebuilt and much expanded in the early 12th century by the Knights Hospitaller of St. John, the castle is a landmark that's impossible to ignore. Of course, neither of the two castles existed when Alexander marched his army along this rocky Mediterranean shore.

Tarsus put me in touch with Cleopatra again. She and Marc Antony met for the first time in Tarsus in 41 B.C. He was a Roman statesman and soldier. She was the Queen of Egypt. History does not tell us how Marc Antony arrived, presumably on a galley from Rome. Cleopatra, as befitted her station (and if history can be believed), arrived with rather more ostentation. She and her minions breezed in from the sea in a flower-bedecked, perfumed ship decorated in gold and silver and powered by a multitude of sweating slaves – hence the need for perfume, I assumed.

What attracted Cleopatra to Marc Antony? He wasn't obviously wealthy, although he did have money and he was high-born. He was powerful, no doubt about that. He was handsome too, and charismatic. Was that enough for a queen, or did politics play a part?

On the other hand, what attracted Marc Antony to Cleopatra? Was it her beauty? Her sex appeal? Her flamboyance? Her wealth? The possibility of enormous regal power? Or was it the woman herself?

Personally, if she really did look like Vivian Leigh, I think… but then, I was a healthy young man and appearances could make my blood surge.

Whatever the motivation, the records show that Cleopatra and Marc Antony hopped into bed together with rather indecent haste. Perhaps lust should be added to the questions above. Even if lust gave the impetus, their *affaire de coeur* was no one-night stand. These two famous lovers kept their passion for each other burning for almost eleven years – until they both committed suicide. What a waste.

Cleopatra cruised on from Tarsus to Ephesus on her elegant royal barge, rowed by ranks of unwilling slaves, a little later in 41 B.C. Marc Antony was with her. Presumably they had taken a short Mediterranean coastal cruise together as a respite from their rumpled bed in Tarsus. Cleopatra's luxury barge deposited her, and Marc Antony, I suppose, at the now silted up harbour. History couldn't tell me whether they were carried by slaves up the gentle incline into Ephesus or whether they did a royal walkabout to shake hands with the locals. Something tells me Cleopatra declined the offer of a walk and insisted on being carried. Marc Antony accepted a lift also, I suspect.

Cleopatra was in a murderous mood. We do know that. There, on the steps of the Temple of Artemis, she had her younger sister assassinated and so disposed of a claimant to the throne of Egypt in typical bloody style. Families can be such a nuisance.

Tarsus was the birthplace of another notable figure – one with higher moral standards than Cleopatra and Marc Antony. A man named Saul, who became the revered apostle Saint Paul, was born in Tarsus in the waning years of the pre-Christian Era.

I stayed in Tarsus for a few days, not so much to learn more of Antony and Cleopatra's sleepover, nor to look for the young Saul. My reason was rather more prosaic. I stayed to help a couple of Italian mountain climbers repair their minibus. In return they gave me a ride by way of Iskenderun – named for Alexander (probably by Alexander) – and across southern Turkey to Iran. They were going all the way to Kathmandu to climb in the Himalaya.

At Iskenderun I left Alexander for a while. He had marched his armies south to Egypt. Cleopatra, nearly 300 years later, had sailed home across the Mediterranean from Turkey to Alexandria. I would get to Egypt and find them there when I was ready.

With the Italians I sped through Gaziantep, Urfa, Diyarbakir and Tatvan. There were places I would have liked to visit on the way, but the winter weather was atrocious, the Italians were in a hurry and I had been told rides were few and far between on the road to Lake Van. I made notes to return some day to spend a few days at Harran, south of Urfa, and to study the stone heads at Nemrut Dagi up towards Malatya.

We kept going east to Van along the south shore of the salty Lake Van, Turkey's largest inland body of water. Wet snow and sleet kept us company and obscured the view. The windshield wipers flapped back and forth hour after hour. Locals warned of bandits between Van and the border but the only ones we met wore uniforms of the Immigration and Customs services for two nations.

We crossed the border into Iran where we wasted many hours on a bitterly cold day while the customs guys examined each item of climbing equipment in the van and assessed its taxable worth. The climbers worried about having to pay a large duty until a supervisor arrived to take charge. He asked a few questions, looked at some of the equipment and pointed to the mountains. The Italians smiled their agreement. The supervisor scribbled a few sentences in one of the climbers's passports and sent us on our way.

Less than an hour later the Italians left me to wait for another ride at Shahpur. They circled the north side of Lake Urmiah to reach Tabriz. From there they planned to take the main road all the way across Iran and via Afghanistan and Pakistan to India and Nepal.

CHAPTER THREE

After leaving the Italians at Shahpur, my peripatetic itinerary took me to Rezaiyeh on the west shore of Lake Urmiah and from there to Hamadan, the ancient Ecbatana, one of the oldest cities in Iran dating back to the 8th century B.C. Once the summer residence of Achaemenian kings, the city fell to Alexander's forces in 330 B.C.

Other than its carpet industry, exotic but far beyond my tiny budget, and the supposed Mausoleum of Esther and Mordecai, Hamadan itself held no interest for me. It was just another stepping-stone on the road to Qum and then the fabulous Isfahan.

I stopped briefly in Qum to view the gold-domed Fatima al-Masumeh Shrine from a distance. The distance was dictated by propriety. Non-Muslims are not welcome at the shrine, or close to it. The gold, or gilded, dome and the tall, slim minarets make it an impressive building, from the little I could see. I walked away to the south, looking for a ride to Isfahan. When Alexander came through this region, long before Fatima's time, he was going north. We passed without knowing it. I was walking. He was riding Bucephalus, his favourite horse.

Alexander the Great. He was my age when he inherited the crown from his father, King Philip. Alexander, at twenty, was a king and a general with his own army. With all that under his control, small wonder that he set off to explore and conquer lands far to the east. I was not so well equipped. I came from humble beginnings. My father wasn't regal material. Far from it. He was a truck driver with no education. Alexander and I had something in common, though, apart from our youthful ages. We were both restless. We were both inquisitive. We were both fearless. And, we both enjoyed a good walk. Even though he had that fine horse named Bucephalus, to keep his tired army in good spirits, Alexander was known to walk with them sometimes through the heat of desert days.

Yeah! Alexander go walking! Yeah! Tinker go walking too!

At Isfahan, as winter began to lose its grip on Iran, I stepped aside from my pursuit of Alexander to explore what many consider to be the most delightful city in Iran. The Isfahanis think so too. They claim – *Isfahan nisf-I-jihan* – Isfahan is half the world. Not exactly a modest appraisal, but an acceptable one. Let me say it: Isfahan is gorgeous! I have never seen such wonderful examples of classic Islamic architectural art. I stayed a week. I felt I could have stayed forever except – Alexander would not let me leave him, yet.

Attempting to describe Isfahan would be like attempting to describe the Mona Lisa to a sightless person. Words cannot do justice to either. The only description of Isfahan must be with the eyes and they cannot share their visions. However, I shall do my best.

Built from an oasis in a fertile valley in the 7th century, Isfahan has survived many incarnations. It was briefly the Persian capital in the 11th century under the Seljuks. The Mongols stayed a while, so did the Timurids. Neither occupiers did much to enhance the city. The Samovids began to improve Isfahan when Shah Abbas the Great came to stay with his court in 1598. Born in Herat, Afghanistan, Shah Abbas chose Isfahan for his capital because of its strategic locations – about as far away as possible from the aggressive Ottoman Empire in the west, and midway between enemies in Mesopotamia to the south-west and Azerbaijan in the north.

Isfahan was my delight. The Persians had the artistic flair to use shades of blue to wondrous effect, none more so than on the Masjid-I-Shah, the Shah's Mosque. Dating from the two hundred years of the Safavid era – 1598 to 1736, the mosque stands at the southern end of the Maidan-I-Shah – the spacious Shah's square. Building commenced on the mosque in 1611. The portal and its two minarets took four years to complete and the mosque itself was under construction for a further fourteen years. Although the entrance portal to the mosque is framed by twin minarets, it does not face the east, as is the custom. Rather, it was built to fit the layout of the Maidan. Behind the entrance, set back from the square, are two more minarets and the massive cupola, correctly orientated towards Mecca.

The entrance portal is a relaxing study in shades of blue, with the lightest blue – a turquoise – again used for the mosque's four minarets and for the exterior of the cupola. To add to the shades of blue are white patterns – geometric in shape – and gold lettering. The overall effect is one of typical artistic Persian grandeur, without ostentation. An Iranian student told me the ceiling of the cupola is decorated in a deep blue inlaid with white and gold floral designs. Sadly, for me, the mosque was not open to non-Muslims at that time.

On the east side of the Maidan, the Sheikh Lutfollah Mosque is smaller than the Masjid-I-Shah yet it is just as richly decorated. It was named in honour of an Islamic scholar from Lebanon who visited Isfahan as a guest of Shah Abbas I and stayed to officiate at the school of theology.

Sheikh Lutfollah Mosque

The Maidan-I-Shah, once a parade ground and a polo-playing field, had survived into the 20th century as a haven of peace festooned with the green of bushes and the vibrant colours of native flowers. From a wooden bench among the gardens I could sit for hours watching the daylight changing colours on the two mosques.

In the evenings, especially the cool of summer evenings, Isfahanis go walking. They stroll in twos and threes along Chahar Bagh, an avenue shaded by plane trees. I went walking too, although on my own. Other walkers smiled at me, nodded their heads in greeting. Some voiced pleasantries. I smiled in return, enjoying my solo promenade.

When I tired of the architectural wonders for a while, I lost myself in the labyrinthine streets of the Great Bazaar in the old town. The narrow covered lanes lined with shops are always busy with buyers from within Isfahan and from outlying villages, plus a few foreign tourists. It is a noisy, vibrant, exotic place to explore at any time of day or evening.

Everything, it seems, is for sale in the bazaar. Grocers, fruit merchants, butchers, hawk their wares. Craftsmen, such as artists, goldsmiths, silversmiths, even shoemakers and shoemenders, show off their skills. I found a street where mechanical parts, mostly from old cars or trucks, shared space with a tire merchant. That was a street for tinkers. I didn't buy anything. I was happy looking at the wares and discussing parts with the vendors. As do many Isfahanis, I sat for long periods at a round table outside a small restaurant sipping tea, munching on sweet biscuits and watching the people passing by.

In one of the squares, where streets meet to share their commerce, I found a stall owned by a young man of superior talent. He worked at painting miniatures using a fine brush and delicate inks. I watched for a while as he transformed a square of parchment, no larger than the palm of my hand, into a perfect rendition of the Masjid-I-Shah. Among his many other offerings were images of Shah Abbas the Great, the various mosques, soldiers on horseback and tiny, vivid scenes of pastoral activity. A strip of silk, or similar material, held a delicate coloured sketch of *Pol-I-Khaju* – the Bridge of Khaju.

A legacy of Shah Abbas II from the mid-17th century, the *Pol-I-Khaju* spans the Zayandehrud River and is a weir as well as a bridge. With 23 arches and two storeys, it reminded me somewhat of pictures I had seen of the London Bridge of the 1600s, but without the shops crowding on each side. *Pol-I-Khaju* is 436 feet from one end to the other and 39 feet in width. Shah Abbas II is believed to have watched the sunset over his city from the central pavilion of the bridge. I understood why he would do so.

Upstream from the Khaju Bridge is an even larger bridge, though not as pleasing to the eye. This one, the Bridge of Allahverdikhan, has 33 arches to support its 175 yards span.

I had the pleasure of meeting an old school pal in Isfahan. I was standing outside the luxurious Shah Abbas Hotel (once an important caravanserai), admiring its façade and wishing I could afford to entertain myself inside, when Richard (not his real name) came out and walked past me without recognition.

"Richard?" I called out. He turned and stared at me, his eyes taking in my travel-stained clothes, my deep tan and the pack on my back. I took off my safari hat and grinned at him.

"I'll race you around the Maidan," I challenged.

He did a sort of double take, then his eyes began to smile and his mouth followed.

"Tinker Taylor?" he asked. "Is that you, Tinker?"

We shook hands. He appeared to be a little flustered, looking about as if he preferred not to be seen with me. I suppose the contrast between my clothing and his immaculate grey suit, white shirt, smart tie, and clean fingernails was a bit obvious.

"What are you doing here?" he asked.

"Oh, I'm following Alexander, you know, the great one from Macedonia," I replied.

"L-l-look," Richard stammered, "I'm off to a meeting. Could you, perhaps, come back about six this evening? We could have a drink in the bar. You do have some clean clothes with you, I assume?"

"Sure, I'll be here," I answered scratching my beard. "I expect I have something presentable to wear in my room."

That afternoon, checking my less than adequate wardrobe, I discovered I had not a clean stitch to wear. Dirty clothes would not do when conversing with the chap once reckoned to be the best dressed young man at our school. I went to the bazaar and bargained for a cotton shirt of pale blue and a pair of faded jeans that sported a label claiming 'Genuine Levis.'

Few Iranian men wore ties, I had noticed. I didn't think I would need one, even in a spiffy Persian bar. I had a wash, trimmed my beard to a reasonable stubble, scrubbed my fingernails until they hurt; then I got dressed. When I walked into the opulent hotel at the appointed hour a doorman in exotic Persian attire directed me to the bar. Richard was already there, seated in a corner facing the door. He was, of course, dressed as an English gentleman. I knew his fingernails would be cleaner than mine, despite all the scrubbing.

Richard had been two years ahead of me at school. We had become friends through a mutual interest in gymnastics. We were team mates there, and we were both members of the senior cross-country team. We often paced each other when in races. I knew he had gone up to Oxford to read history after leaving school.

He had the decency to ignore my casual clothing as we shook hands. He motioned me to sit at his right. We ordered drinks. I had a Scotch and soda. He chose a very British gin and tonic.

"What are you doing here in Iran, Rich?" I opened the conversation.

"Oh, I'm attached to the consular section of our embassy in Teheran," he explained. "I'm just taking a few days leave in Isfahan."

From the way he kept glancing at the door, I could tell he was watching for someone. I wondered if he had been recruited by MI5, or 6, or one of the other numbers, as a spy while at Oxford. So I asked him.

"Are you a spy?"

He looked shocked; then he gave a weak smile, took a sip of his drink, shook his head and said, "No, no, Tinker. Nothing like that, I assure you."

I didn't believe him.

We talked about my many months on the road and my reasons for the journey. We talked about our school days. We laughed at the time he tripped and fell during the last 100 yards of a six-mile cross-country race he was leading. He had been quite piqued at the time because I jumped over him as he rolled on the ground and I won the race. We talked about my interest in the wonders of antiquity. We were laughing about one of our teachers, a humourless man who tried to teach us chemistry, when his face changed. Richard wasn't looking at me. He was looking at the door to the foyer where a young man stood watching us. He had one hand in his jacket pocket, otherwise he looked alert. Richard waved and motioned him to join us.

Richard was a handsome man, slim and just over six feet tall, with dark hair and brown eyes. The newcomer would have ranked higher on the beauty scale. Blond hair, blue eyes, pale skin, same build as Richard and, like him, clean and manicured nails.

"Tinker, I'd like you to meet my friend, Sebastian," Richard made the introductions adding, "Tinker is an old school friend. We used to race against each other."

Sebastian was dressed in impeccable style like Richard. He had a firm handshake, but his voice betrayed him. There was an accent, which

I soon discovered was Swedish, and there was something else. He was a flight steward with SAS, the Scandinavian airline, based in Copenhagen and working the Middle East route. The light clicked on in my brain when he smiled at Richard. Hmm, that's interesting, I thought. I didn't know that about Richard.

I enjoyed the two hours I spent conversing with them in the bar, though I declined the invitation to join them for dinner. They obviously wanted to be alone, in spite of their courtesy. Richard insisted on paying the bar bill, which I appreciated. I shook hands with both of them as I left to go in search of an eatery I could afford: one with cheap chelo kebabs – rice and ground meat with onions –accompanied by hot, strong tea. Dinner was excellent. Instead of reading, as I normally did while eating, I thought about Richard, not about his sexuality, but about his clandestine work, convinced he was a British government spy.

I left the Isfahan of Shah Abbas, my eyes and mind sated with architectural wonders and of dominant blue tiles reaching to the complimentary azure of the cloudless sky. Ahead of me, a few hours down the road, and centuries behind, was the real reason I had ventured so far east from Europe.

TINKER GO WALKING!

CHAPTER FOUR

Persepolis was my goal. The city planned and started by Darius I and built by successive rulers over a span of fifty years was to be the eastern extent of my quest. I arrived from Isfahan in the cab of a truck laden with over-ripe animal skins. The driver let me out opposite Persepolis on a fine spring day and waved goodbye with a shouted comment.

"Is old place now, but good!"

With my pack on my back, I stood in awe in front of the monument I had travelled so far to explore. My research had not prepared me for the dramatic scale of the site. The walls stretched far to my left and far to my right. Stately columns reached to the sky. I imagined the thousands of labourers and artisans who toiled for so many years to create this architectural wonder. I imagined the Gods watching and approving. I imagined Alexander's restless army of untold thousands camped on the plain in front of Persepolis. It was almost too much to take in. Persepolis had been a thriving city of intelligent people while Britons were still living in caves, wearing dirty animal skins, and grunting to communicate. That knowledge made me feel insignificant. And that was before I had walked its spacious terrace.

In slow motion, step by step, I walked up the two levels of the wide entrance stairway to the Gate of Xerxes. The west door is protected by two pairs of huge stone bulls. The east door is guarded by two winged bulls adorned with human heads. Who modelled for those? I wondered

After passing the bovine guardians, I stood in the courtyard on the main terrace, 1475 feet long and 985 feet wide, and looked around me. I left my backpack under one of the winged bulls and asked for its protection. I preferred to be unencumbered for a few hours.

Across the courtyard I came to the north steps, one of the glorious Apadana stairways leading to the royal audience hall. There I bowed my head in reverence to Darius, to Xerxes, Artaxerxes I and III, and to Cyrus, and I shook my finger in admonishment at an inebriated Alexander for playing with matches.

The Apadana stairways of black marble are exquisitely decorated with lines of Persian, Mede and Susian guards and their armies. Then there are seventeen groups from different nations bearing tributes for Darius. Among them are delegates from Armenia, Babylon, Lydia, Sogdiana, Cappadocia, Bactria, Egypt, Scythia, Ionia and Susa. Their gifts run the gamut from horses, camels, precious metals, and fabrics, to weapons of war. The gifts are the best each land has to offer for the all-powerful king of the Persian Empire in the 6th century B.C.

Originally decorated in vibrant colours, which must have been hard on the eyes when combined with bright sunshine and a clear blue sky, time and winds have scoured off the paint and left the black marble bas-reliefs. Not so colourful, but much more impressive to my mind.

Darius was never afraid to express his ego or to extol the virtues of others in his kingdom. He said, "Ahuramazda, the greatest of the gods created me, made me king, bestowed upon me this great kingdom, possessed of good horses, possessed of good men."

Darius didn't need to build Persepolis. He already had a splendid capital at Susa, of which he exclaimed, "At Susa a very excellent work was ordered, a very excellent work was brought to completion."

As I said, he didn't need to build Persepolis yet I, among many others, am so pleased he did. Persepolis is a work of exceptional art. Darius and his successors must have hired the best minds and hands in the realm to have designed and built this glorious new capital. Untold numbers of mathematicians, engineers, sculptors, stone cutters, artists, artisans and painters turned their skills to creating an elegant city that would last for thousands of years.

Two griffons with hooked beaks and pointed ears appear to have no purpose other than to show off the sculptor's skill. From one angle a griffon appears to be about to devour a distant tall column, an optical illusion created by distance and angles. Clever mathematics from antiquity.

Looking up at the face of the mountain backdrop I could see two large cruciform openings: the tombs of Artaxerxes II and his successor, Artaxerxes III. They've been up there in silent repose since long before Alexander arrived.

Laid out on a natural terrace, expanded with additional flat paving stones by an army of artisans to 125,000 square yards, Persepolis was the brainchild of King Darius I. When construction began about 512 B.C.

the natural terrace set against a mountain wall was a perfect partial foundation for the city Darius envisioned.

I walked the vast terrace, stopping to admire carvings older than many civilizations. In my excitement, I moved at speed from one wonder to the next. My first day at Persepolis left me exhausted in mind and body. I stayed that night on a marble slab, like a sacrifice waiting for the inevitable execution. Hardly able to sleep, I reflected that Persepolis had already been inhabited for over 200 years when Alexander and his army arrived. When I walked those gracious lanes between towering pillars of black marble and stone carvings of exotic creatures. Persepolis had stood in dignified supremacy on the site for almost 2,500 years. How much more graceful would Persepolis have been today if King Phillip's son could have curbed his arsonist tendencies?

None of the scholars I have read can offer adequate explanations as to why Alexander burned Persepolis. They all agree that he was drunk, as were most of his men and the camp followers. Did someone, maybe a woman named Thais, whisper in his ear, "Hey, Alex, have another drink and then let's start a fire and watch the palace burn."

Whatever the reason, whether it was drunkenness, political expediency, the conquering spirit, or just plain vandalism, Persepolis burned and the responsibility for the desecration fell on Alexander's royal shoulders. Leaving the ruins smouldering in his wake, and suffering from a hangover, he marched the army on to continue his campaign to dominate the Persian empire. Alexander didn't know it, but a young woman was waiting for him at her father's palace a few months ahead.

Alexander and his army continued far to the east, through Afghanistan and into what is now Pakistan. I considered following their trail but that would have taken me so much further away from my ultimate goal, which was Timbuktu, in West Africa, on the south side of the Sahara. I let Alexander go on without me. He had a woman to meet.

Roxanne was the daughter of Oxyartes of Bactria, north-east from Persepolis, far away beyond the Hindu Kush mountains. Roxanne was sixteen years-old when Alexander saw her dancing in his honour at Bactria. She was said by many historians to have been the most beautiful woman in the world. (Now, where have I heard that before?) I'm sure she was beautiful. All paintings and sculptured renditions of her show Roxanne to be so.

Alexander was captivated by Roxanne. He married her in 327 B.C. although I could never find out where the ceremony took place, only that it must have been somewhere in Bactria. Legend says that Roxanne knew Alexander had a mistress, or maybe another wife. She married him anyway, bore him a son and was with him when he died of a fever at Babylon in 323 B.C.

Seated on a smooth marble stone, I closed my eyes and imagined Alexander walking through what was left of Persepolis on his return visit about 324 B.C., the lovely Roxanne by his side. Did he explain to her his reasons for the deliberate conflagration? Did she get mad at him for the destruction of a royal edifice? Did she have the foresight to complain that his wanton abuse would go down in history and damage the illustrious family name?

On that note – the wanton abuse of history – my guidebook to the Persepolis region recommended spending half a day visiting Persepolis and Naqsh-I-Rustam. The other half of the day, it said, could be used to visit the Tomb of Cyrus the Great at Pasargadae. One day to see all those treasures? Only one day? The very idea horrified me. That was wanton abuse of historical time.

A coachload of British tourists arrived at Persepolis the following day in the early afternoon. They poured out of their bus like ants on a foraging mission and scurried in chattering groups all over the hallowed stones. They stayed no more than three hours, as the guidebooks recommended, while I retreated to a shady spot against the cliff. They left as they had arrived, chattering about the wonders they had just seen, about their sons and their daughters. I caught the tail end of a conversation.

"Did you know Alexander the Great lived here?"

"He did? When was that? I thought he was Greek, or something."

The voices faded onto the bus with their owners. The bus drove away in search of the next mini cultural adventure for its passengers.

About an hour's walk north of Persepolis, at my pace, is the necropolis of Naqsh-I-Rustam. Four Achaemenian kings were buried in cruciform tombs carved into the sheer face of Mount Kuh-I-Hosayn. The tombs are identical to each other in pattern and style, and identical to the two at Persepolis. They house the remains of Darius I & II, Xerxes I and Artaxerxes I who reigned in turns over Persia between 521 BC and 405 BC.

I'm fascinated by grave sites, perhaps because I shall have one of my own at some time in the distant future, I hope. I have seen many in my travels, though none so overpowering as those at Naqsh-I-Rustam. The tombs are towering cross-shaped examples of the grave-digger's art.

The upper horizontal of each tomb holds a bas-relief depicting citizens holding a throne aloft with the king seated, all watched over by Ahura-Mazda, Lord of Wisdom, the highest ranked of the Zoroastrian gods. Ahura-Mazda stands watch over many elaborate historic sites of ancient Persia, including Persepolis.

A series of eight Sassanian reliefs have been carved into the rock face below the tombs. Most of the bas-reliefs I saw in Iran showed wonderful touches of artistic flair. At nearby Naqsh-I-Rajab is one of my favourites. An unknown beardless dignitary holds up his right hand, presumably as a sign of respect or homage, to the royal family of Ardashir I in the next relief.

Forty miles south of Persepolis is Shiraz, the city of poetry. I covered the distance in a truck in about an hour and stayed long enough to read ancient Persian poems at the tombs of Saadi and Hafez. Both interred in lovely gardens of lawns, flowers, bushes and trees, they are peaceful invitations to stop and rest. Hafez wrote, "*I wish I could show you when you are lonely or in darkness the astonishing light of your own being.*" Those lovely words carried me forward, as did the encouraging, "*The place you are right now God circled on a map for you.*"

I left Shiraz with a new understanding of Persian poetry and increased admiration for Saadi, Hafez and Omar Khayyam – the Poetic Persians of a gentler era. I also had a long wooden staff, exactly my height, which I had found in a ditch.

Turning west again, I hitched a ride on a truck to the oil fields at Ahwaz. I had hoped to get a job for a few weeks, but no one was hiring. I caught another truck heading north to Kermanshah. A few miles towards the border with Iraq, an hour north of Ahwaz, stands the site that was once the proud city of Susa. The few parts of Susa that have been uncovered are scattered across four hills of the Zagros Mountains. A fellow traveler had told me in Isfahan that the side excursion was not worth the effort. I took him at his word.

There was one detour that I was determined to take. About 25 miles from Kermanshah, high up on a rock face, reachable only by a long extension ladder, or a complicated set of scaffolding, neither of which was available, is a bas-relief commissioned by Darius the Great. Known locally as Bisitun Rock, the relief commemorates Darius's victory over those who had plotted to oust him from power. Like all other visitors, I had to stand far back and incline my head to see the carvings.

Closer to Kermanshah, only an hour's walk for me, is Taq-I-Bostan. A large pond stands beside shady trees in a pleasant garden at the foot of a cliff which contains 4th century Sassanian bas-reliefs, similar in style and execution to those at Naqsh-I-Rustam. These, however, are more descriptive, telling tales from a series of events. The hunting scenes, with Ardashir II victorious over a wild boar from a boat in the marshes, a stag, and an elephant, set against a background of tall reeds, are spectacular.

At Taq-I-Bostan I saw what appeared to be an unintentional symbol for the future. A little Iranian boy lay asleep on a blanket beneath a bas-relief of the tree of life. I whispered my wishes for a peaceful life and left him there. I have often thought of him since and wondered how he fared.

A succession of short rides carried me north from the Kermanshah area, out of Iran and into Iraq. My journal reminds me it took all of one day to persuade the Iraqi border guards to let me into their country – even though I had the correct visa and was using an easily recognizable United Kingdom passport. Thanks to their suspicions I spent an uncomfortable night at the border, sleeping fitfully with my head on my backpack, and a few hours doing the same the next day, before a friendly Iraqi took me to Mosul.

The smell of oil thickened the air in Mosul, as it had done for me ever since reaching the first oil fields south of Ahwaz. I didn't stop long. My driver directed me to a safe place to wait for my next ride. A passing donkey ignored my greeting, as did its owner. A couple of stray dogs sniffed at me without showing further interest. A few small boys watched me from a distance but did not come close enough to start a conversation or play with a ball. I waited a few hours, lost in my own thoughts getting hotter and getting dustier, dirtier and absorbing more oil through my pores minute by minute until a car pulled up and the passenger asked me in French where I was going. I told them to Palmyra in Syria. The passenger said they could take me as far as Raqqa, also in Syria.

When I crossed borders with other travelers from east to west I was always nervous in case any of them had contraband, such as illicit drugs, with them. On those occasions I never let my backpack out of my sight, usually carrying it on my knees. It was a sensible precaution. I had heard too many tales of innocent nomads such as me being caught with someone else's smuggling tucked into their bags.

The Tree of Life

At Raqqa the Frenchmen continued west. I walked south across a bridge over the Euphrates and headed in the direction of Palmyra, the Tadmor of Biblical times. I had then crossed the Tigris and the Euphrates rivers twice each. Once each in Turkey and then in Iraq for the Tigris and in Syria for the Euphrates. They are both big rivers. Both were huge challenges for Alexander and his army, but not insurmountable.

On the south side of the river, opposite Raqqa, I waited for an hour before a truck of doubtful provenance belched into view. I waved to the driver and pointed south, he slowed and beckoned me to the truck. He didn't stop. The truck rolled forward at a walking pace. I opened the passenger door, heaved in my pack, pushed my staff behind the two seats and hauled myself after it.

"Where you goin', mate?" the driver asked in what sounded like a parody of a Cockney accent.

"Palmyra," I answered.

"Okay, old son," he said with a stubbly grin. "I'm going to Homs. I'll drop you off on the way."

I looked at him, a smile spreading from ear to ear. "Old Son?" That was a typical rural London form of address from one man to another.

"Where did you learn your English?" I asked.

"I worked for the old raff at Habbaniya for a few years," he said, his grin as wide as my smile. "I was a gardener there."

"You mean the RAF station in Iraq?"

"Yeah. Bloody good job that was, mate, apart from the screaming of those Vampire jets. I'm from Baghdad. Name's Pete."

"Pete? An Iraqi named Pete?" I burst out laughing. He laughed with me.

"Yeah. Actually, I'm Syrian but I grew up in Baghdad. The raff bloke I worked for at Habbaniya thought there was too many of us Mohammeds on the base so he looks me up and down and says, 'You look like a Pete to me', so Pete I've been ever since."

As we rattled in a general south-easterly direction on a vague track, we chatted about the good old days when the British forces in Iraq employed local labour. Pete admitted he missed that life.

"I did well. I made good money, but gardening don't pay much these days. Now I drive this old crate for my brother-in-law. I've got a load of carpets back there," he indicated over his shoulder. "Gotta deliver them to Homs, then down to Damascus. That's where I live now. Got a wife and kids there."

Before Pete left me at Palmyra he wrote his address in Arabic script on a piece of paper. "Come and visit me in Damascus," he said. "I'll show you around."

Oh, how my mind wrapped itself around the glory that was Palmyra. Standing at the crossroads of civilizations in the 1st and 2nd centuries, its perfect colonnades and the remains of its massive walls had begun to dominate the western horizon as Pete and I drew near.
Persepolis was the most graceful example of an ancient city I had yet seen. Palmyra came a close second.

CHAPTER FIVE

An overnight stop for camel caravans trading between distant cities and nations, Palmyra in some form or other has been in existence in the middle of the Syrian desert since before 2,000 BC.

I waved goodbye to Pete and his smelly old truck as it spluttered into the distance. Taking a deep breath of clean, warm air, I walked into history. There were no other humans in sight, only a single donkey asleep in the shadow of a wall built before Alexander came here.

The shadows were lengthening as I walked along the colonnade. I had a chat with the donkey about accommodation. He opened one eye and then the other and looked at me. He didn't want to talk. He shook his head to discourage a few flies and closed his eyes again. His head settled on a flagstone and he drifted into sleep. Assuming he wouldn't mind warm-blooded company, I spread my sleeping bag nearby.

Night fell as I lit a small fire of twigs and brewed a mug of tea. Leaning my back against a pillar I dipped stale bread in the tea and ate that with a piece of cheese that had seen fresher days. I slept well, dreamless, only waking for a few seconds late at night when the donkey wandered away on business of its own.

Daybreak saw me walking through the monumental arch and along the considerable length of the colonnade, touching the pillars and noting their surroundings. When the colonnade was completed there were three sections with a total of 1,500 tall white columns along its length of roughly three-quarters of a mile. Time, weather and a certain amount of vandalism have reduced the columns in height in some cases and in significant numbers in others. Above the city, perched on top of a steep hill, is the castle of Fakhr-al-Din al-Ma'ani – said to have been built by Saracens in the 13th century. Obviously, the castle had nothing to do with Palmyra, apart from its dominant hilltop position.

I had to limit my architectural studies to the early morning and the late afternoon. During the middle hours of the day, when the desert sun seared the land, I relaxed on my sleeping bag in the shade of a wall to avoid the fierce heat. There I searched my Bible for a reference to Tadmor. I had added a pocket Bible to my belongings before leaving home in the belief that the history of the lands through which I was travelling might make additional sense when combined with some religious studies.

In part that was true. Some of the stories in my Bible came alive for me as I moved from town to town and from country to country. Many remained obscure, more obviously fables than historical fact. I found Tadmor in the *Second Book of Chronicles 8:4*. "And Solomon went to Hamath-zobah, and prevailed against it. And he built Tadmor in the wilderness, and all the store cities, which he built in Hamath."

Tadmor, therefore, was built, or perhaps re-built, by King Solomon, the Israelite. Solomon, he of the wise decisions, and known to the possibly exotic queen of the land called Sheba, ruled Israel after King David, between 970 and 931 BC. The Hamath mentioned in *Chronicles* is now the city of Hama, in Syria, between Homs and Damascus.

Archaeologists believe that the Palmyra ruins spread over the desert – the columns, the avenues, the triumphal arches, the tombs, the temples, the administrative buildings and the houses of the citizens of Palmyra – date not from Solomon's time, but from close to 1,000 years later. Solomon's Tadmor, lies hidden near or under the present ruins of Palmyra. It seems there is a gap of nearly ten centuries in Tadmor/Palmyra's history. How could that be? I wondered. What architectural treasures might one day be discovered beneath Palmyra's stone and sand?

I could not begin to imagine what happened at Tadmor during any of those lost years. I could not imagine an historical void of 10,000 decades. My only points of reference were the erratic building blocks of broken objects around me, and they could not speak. Sitting in the middle of the colonnade I could picture life at the beginning of the

Christian era. The donkey gave me a clue and a pair of camels in the distance added to the thought.

In its heyday this sprawling city would have played host to merchant caravans from as far away as Bactria, India, Arabia, African lands bordering the Nile, plus Rome, Greece, and perhaps from Byzantium. The traders would have arrived with Bactrian camels, dromedaries, donkeys and horses. Some would have had dogs with them; some with wives and children, for the journey was arduous.

Those long-distance salesmen wore colourful robes and turbans of many shapes. They were short. They were tall. They were thin and they were fat. Some had braided beards and others were clean-shaven. Their skins were all shades from pinkish white to the black of Nubia. Their eyes ran the spectrum from slate grey to obsidian black. They journeyed prepared to defend their goods from any attack. To that end they were armed with straight swords, with curved scimitars, with daggers and with lances: their bladed weapons forged from iron and from bronze.

They spoke, as the saying goes, in a Babel of tongues, some loud, some quiet. Most conversed with the authority of the hardened traveller and the determined merchant. Despite their linguistic differences, they understood each other, somehow, and they understood those who wished to buy from them. Trade has many languages.

Palmyra, one of the most important trading centres on the caravan trails between Europe and India, has experienced a stormy history. From just before the beginning of the Christian era Palmyra was ruled by the Romans for over 200 hundred years. Then the Persians took the city away from them. The Persians managed to hold on to Palmyra for a handful of decades until Rome's Emperor Aurelian said, "Our turn again." His army wrecked the city in the late 3rd century. Then they rebuilt it. Rome kept control of Palmyra for the next 400 years, until various caliphates of the Arab world became dominant. If only the missing 1,000 years could be resurrected, how much more excitement would Palmyra's past reveal?

TINKER GO WALKING!

CHAPTER SIX

I managed to get myself in trouble leaving Palmyra. I was walking, as usual, out to the east-west gravel road where Pete had dropped me two days before. A car stopped and the Arab driver offered me a ride. The car was going in the right direction. I poked my staff between bags on the roof-rack and squeezed myself into the back seat with three other Arabs, thinking all was well. At a small village called Al Husn, the car stopped. We all bailed out. My fellow passengers handed over a few paper notes to the driver as I retrieved my staff from the roof and pack from the trunk and hefted it onto my back. The driver snapped his fingers at me and rubbed thumb and forefinger together. He said something I didn't understand.

"Floos," he said, raising his voice. I didn't know the word, but the sign was obvious. Give me money!

That was difficult because I didn't have any, other than a few coins. The driver got mad. Police arrived. Someone explained to me that I had been in a car, not a taxi exactly, licensed to charge passengers. Ooops, I thought. What now?

A policeman ushered me across the road to a small police station, for my safety, I suspect. I sat on my pack in a corner while a debate raged as to what to do with me. I understood jail was one possibility. Deportation another.

"Where are you going?" an interpreter asked.

"Tel Kalakh; then Damascus," I replied, "after that to Beirut."

A flurry of words that constituted a conversation and an argument passed at high speed between the interpreter and the police, both assisted and impeded by a crowd of locals who had squeezed in to share the occasion. One point was obvious, some wanted me lynched, others would have had me jailed. I failed to appreciate either possibility. A voice of reason from a man dressed in a western suit suggested a compromise. Pointing to a page in my passport, he said, "He has a visa for Lebanon. I'll drive him to the border. He can cross there."

I didn't want to go to Lebanon yet. I wanted to roam the old crusader castle at Tel Kalakh. I wanted to explore Damascus. I stood and pleaded my case. My English was better than theirs. Their Arabic was better than mine – much better. I lost the case.

"You go Lebanon," a finger in my chest emphasized the decision.

And so, despite my protestations, the interpreter in the western suit ushered me out of the police station, into the back of a car and, a short while later, stopped at a Syrian border post. By this time the midnight hour was fast approaching. The Syrians stamped my passport and wrote something rude across my visa in Arabic.

My driver pointed into the future. "Lebanon," he said. "You go to Beirut."

My visa for Syria had been cancelled. I had been deported to the next country on my list. The way ahead was dark. I walked south, taking short steps down the middle of the road into the Stygian unknown. Annoyed at my unceremonious departure from Syria, because I hadn't seen as much of the country as I wanted, I complained aloud to myself. I had not yet been to Damascus, the oldest continually inhabited city in the world. Faint laughter drifted to me from ahead, breaking into my soliloquy. A light flashed. A voice called in French, "M'sieu. Venez ici." Come here!

The Lebanese border guards proved to be highly amused at my predicament. Obviously, they didn't get much entertainment at that outpost. They made me tell my story, so much easier in French, and laughed at every nuance. They slapped me on the back. They fed me a meal of bread, dates, meat and rice. They gave me tea. And they persuaded me to sleep on their floor until daybreak. I was happy to comply.

In the morning, with the sun not yet visible in the east, I had more tea, and an omelet. The guards refused to accept any payment but allowed me to clean the dishes at an outdoor water pump. They suggested I wait. There would be a car or a truck, or a bus, or some form of transport along some time that day. I thanked them, pointed to my feet and said, "Non, merci. Tinker en marche à pied." And away I went, on foot, with a comfortable feeling in my belly, south along the coast road towards Tarabulus and onwards to Beirut.

Away from the border I found a path down to the sea. Not having had a good all-over wash for many days, I stripped naked and went for a swim. The water felt good as I rubbed the grime of desert travel from my pores. I took the opportunity to wash some underwear and socks, plus a shirt, all of which dried in minutes when I spread them on the ground under the hot sun. I stayed there the rest of the day and slept on the beach that night. I was clean and I was comfortable.

An Irish girl I met at the hostel in Beirut a few days later told me of a way to earn easy money in Lebanon. "Give a pint of blood at the university hospital," she advised. "I did and they pay five pounds sterling."

Five pounds was about twenty dollars. I gave blood, rested for a while, accepted a cup of tea and a sweet biscuit, and collected my money. I could live on twenty dollars for a few weeks.

Lebanon has long been referred to as the Switzerland of the Middle East. Beirut is a cosmopolitan multi-lingual city where Christians and Muslims live close together and in harmony. The narrow country has wonderful beaches on one side and forested hills on the other where the skiing is considered world class. Also up there in the hills is the 2,000 years-old Phoenician and Roman treasure named Ba'albek looking over the Beqaa Valley.

Less than three hours from Beirut on a narrow winding road with little traffic, it feels much more isolated. Like Palmyra, Ba'albek is far enough away from urban areas to have resisted the encroachment of group tourism – so far. I shared expenses with eight other nomadic Europeans in a passenger van driven by an exuberant Belgian from Ghent who had no idea of the rules of the road.

I closed my eyes and prayed as he raced around blind corners, somehow squeezed past a truck where there was not enough room and scraped the hillside too many times. By the time we reached Ba'albek we were all silently screaming in fear. And we still had to go downhill with him at the end of the day.

Thanks to the crazy Belgian driver, my nerves were not in good shape as I roamed Ba'albek. At most archaeological sites I explored with my mind focused on the location. At Ba'albek I found my mind drifting, worrying about the ride back to sea level. I knew there was no alternative. I either had to walk, pay an expensive bus fare – if a bus should come along – or walk all the way. I tried to concentrate on the present and enjoy the wonders of the site. To bolster my spirits I reminded myself that Alexander came here as well, though not with a demented Belgian driving his chariot.

The temples attracted religious pilgrims to worship their favourite deities: Jupiter, Venus, Bacchus, and Mercury. Thinking of the coming journey, I wondered which one I should pray to. Should it be Jupiter? No, a god who sent thunder and lightning would not do. What about Venus? Again, no. Attractive though she was, a goddess of love could not help me. My feelings for the Belgian driver were quite the opposite. Bacchus was definitely out of the equation. So, Mercury it had to be. I decided his winged feet would be perfect to help me escape extreme danger. Mercury received my prayers that day and, as I realized later, he looked after me.

The journey downhill was a repetition of the drive to Ba'albek, only much faster. Needless to say, I was terrified that my young life might come to an end on a Lebanese hillside. I stepped out of the van in Beirut, just off the Corniche. My legs were shaking, and my chest felt too tight for normal breathing. That night I had a light meal and then went to bed. My dreams became nightmares as I re-lived the Belgian's bravado behind the steering wheel.

Undeterred by my ignominious early departure from Syria, I obtained a new visa by nefarious means in the Lebanese capital (yes, it was a forgery and it cost me five of my blood dollars). With dark ink spilled over my original Syrian visa in my passport, I went back a couple of weeks later, taking a ramshackle bus through a different border crossing from Beirut to Damascus. None of the border officials noticed the old visa or the new artwork.

Damascus. How that name resonated in my head. A city that had grown under the influence of Hellenistic, Roman, Byzantine and Islamic cultures. A city that had existed for over 10,000 years – not in its present form, of course, but archaeological excavations have shown that there has been a settlement on the site for that length of time. And yet, according to my Bible: *Isaiah 17: 1-2*, "Damascus will cease to be a city and will become a heap of ruins." So far that has not happened and the oracle who spoke those words did not suggest a date for the destruction. May his dire predictions continue to be incorrect.

General Parmenion, one of Alexander's senior soldiers took Damascus in about 334 B.C. It is possible that Alexander came here as well, on his way from Egypt to what would become the battle of Gaugamela. Damascus hiccupped and lived on. The city was already ancient when Jesus Christ and his Apostles walked this much-contested land. One of its most significant visitors was Saul of Tarsus – an angry man determined to wipe out Christianity – all by himself if necessary. He beat Christians. He had them imprisoned. He had them executed. He was a relentless warrior against Christianity. All that changed on the road to Damascus from Jerusalem.

Not far from Damascus Saul was struck by a blazing light, so intense that it knocked him to the ground. Legend says that Saul then heard the voice of Jesus Christ reproving him for his campaign against Christians.

"Saul. Saul. Why do you persecute me?" the voice asked.

When Saul opened his eyes, he found he was blind and had to be led into Damascus. There, Saul resided in the house of Judas on the street called Straight. He remained blind for three days, until Ananias converted him to the Christian faith and baptized him. Saul then regained his sight, no doubt with a huge sigh of relief, and became Paul the Apostle. Eventually the most famous of the twelve.

The street called Straight runs east and west for about a mile across the middle of the city. Although I walked the length of the street on both sides, I saw no sign that Saul or Paul had once lived there. The only interesting reminder of Greco-Roman times is the entry arch over that street called Straight.

As with most Middle Eastern cities, from Turkey to Iran and far beyond, the covered markets with narrow streets crowded with traders and buyers were highlights for me. Damascus was no different. I'm not a great fan of crowds, and yet I love those markets. I can immerse myself in their mysterious depths for days at a time.

I went looking for Pete, my friendly Syrian Iraqi truck driver. I found the apartments where he lived and I met one of his children, or so the boy said, only to discover Pete was away again.

"He's gone to Raqqa," his young son told me.

"Tell him Tinker came to say hello, will you, please?"

The boy nodded, smiled, and pocketed the few coins I gave him. I caught the next rattling bus back to Beirut. There I sold another pint of my precious blood and pocketed the payment. I wanted to go to Israel but the only legal way to do so at that time was to fly in as all land borders were closed. That meant by air to Cyprus and then again by air to Tel Aviv. Air travel was far beyond my miserable resources. I didn't have that much blood in me. Israel would have to wait for a few years.

I stayed a few more days in Beirut because I liked the city so much. I helped repair a truck and a car and earned a few more dollars. I also spent many hours on the Corniche, staring out over the Mediterranean and thinking about my journey so far.

I had travelled thousands of miles following ancient trails trod by warriors, statesmen and philosophers. I had explored Biblical cities and cities older than time. I had walked in the footsteps of Alexander, of Marc Antony, of St. Paul, perhaps of Jesus himself, and, assuming she did walk occasionally instead of being carried by slaves, I had walked in Cleopatra's tiny footprints. Those people, places and historic events had been my university.

I had also aged a year. In that twelve months I had grown physically stronger; mentally more agile. I had devoured knowledge and experiences with an appetite that seemed inexhaustible. I was more confidant. I was happy, but hungry for more. It was time to leave the eastern lands of Alexander and of the Bible. It was time for something completely different.

Instead of spending money I didn't have to visit Israel, I travelled deck class on a small Greek freighter for an overnight voyage from Beirut to Port Said. A wealthy young American who introduced himself as Dennis from Pittsburgh sailed with me. He had a car with him, rented from Paris, he explained. The car spent the voyage perched on a cargo hatch. Dennis had the luxury of a berth below decks. I slept in the open. I fell asleep happy on a gently rolling Mediterranean Sea.

Dennis was travelling solo for six months without any particular itinerary, all expenses paid by his father as a gift for earning a degree in some discipline or other from Cornell University in New York State. He joined me on deck in the morning and offered me a free ride to Cairo, which I was more than happy to accept. After a lazy day at sea our freighter threaded a meandering course between the ships waiting at anchor to enter the Suez Canal. We docked at Port Said in the early evening. The smells born on the breeze circling the harbour intoxicated me with their spices, their mystery. The smells said it all: This is Africa!

TINKER GO WALKING!

TINKER GO WALKING!

PART 3

AFRICA

TINKER GO WALKING!

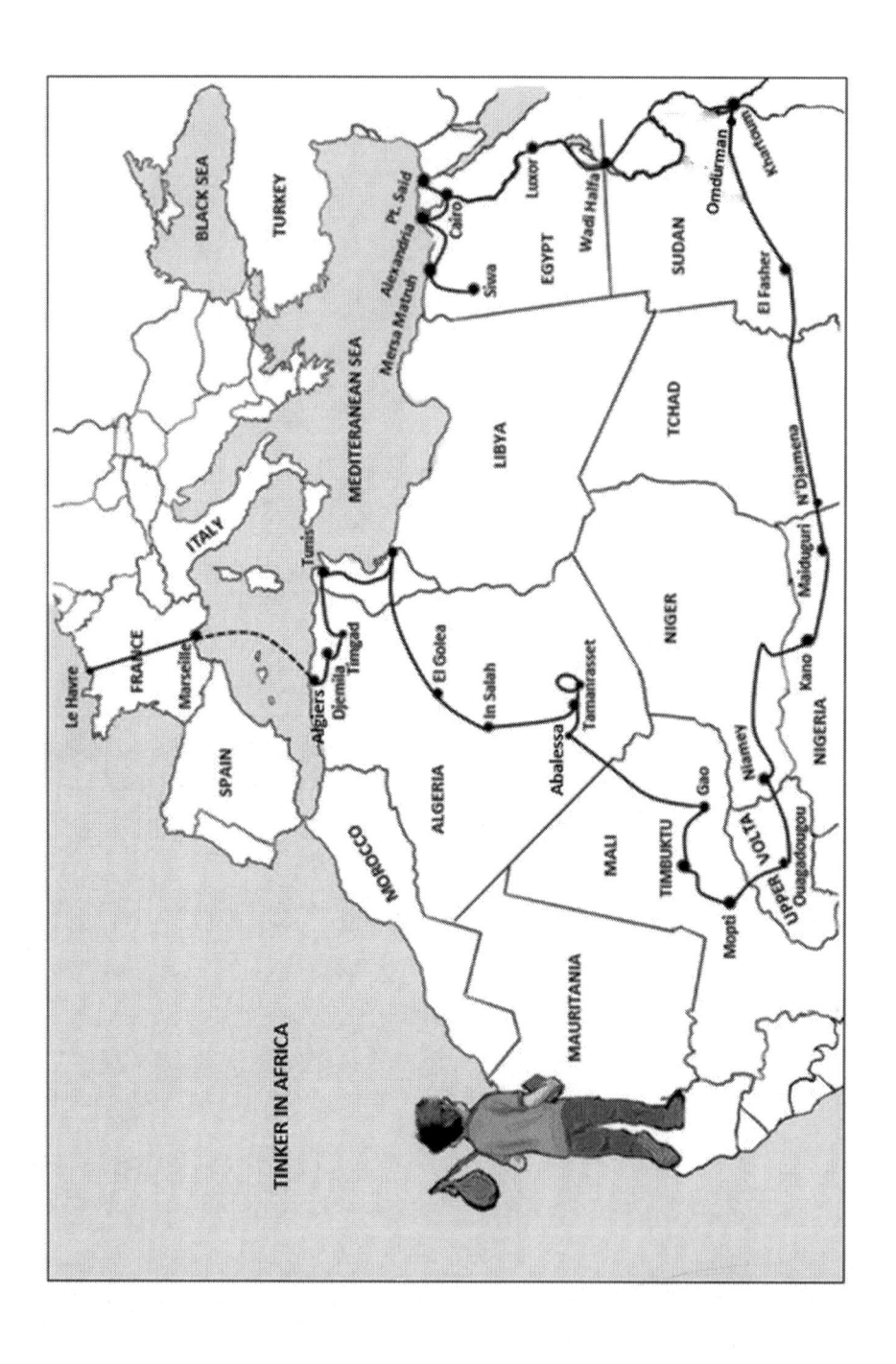

TINKER IN AFRICA

CHAPTER SEVEN

Hello, Port Said. Hello, Egypt. Hello, Africa. I stepped ashore from a wooden gangway onto a concrete dock. The concrete was the same as that used all over the world. The air borne on the casual breeze was warm – a different warm. The sounds were different. The sights were different. The people were different. Herodotus, the acknowledged Father of History, and himself a great traveller, called Egypt 'the gift of the Nile.' I was excited to be there. I looked around me in wonder and said again, "Hello, Egypt! Hello, Africa!"

I watched from the dockside with Dennis as his rented Renault was lifted off the ship in heavy canvas slings. A crane deposited the car on the dock with hardly a bump. He had a reservation in a small hotel for that night and planned to drive to Cairo next morning. I slept in his Renault, as a sort of night-watchman. In return, he would drive me to Cairo. A good trade off.

We drove along the west bank of the Suez Canal for the first half of its 120 miles. On the east side was Asia, where I had just come from, beginning with arid desert lands. On the west side, where we were, was Africa – the fertile silt-laden delta of the fabled Nile River – with another vast desert beyond that. In between was the canal, historic in its own way.

It's possible that the pharaohs had a narrow canal dug between the

Red Sea and the Nile for their exclusive use. No one knows that for sure as there are no records and, if the canal did exist, it has long been filled in by desert sands. Napoleon Bonaparte considered the idea of a canal linking the Mediterranean Sea with the Red Sea after he conquered Egypt in 1798. Busy, no doubt, with his army, he sent the wrong men to survey the land at both ends of the isthmus. Their measurements suggested there was a 30 feet difference in height between the two seas. A canal, therefore, would need expensive locks to avoid flooding the Nile delta. Napoleon dropped the idea. Nearly 50 years passed before more careful surveying proved there was, in fact, no difference in level between the two seas.

Another Frenchman, Ferdinand de Lesseps, engineered the present canal, taking 10 years to carve a wide trench through the Sinai Peninsula. For most of that time he fought a war of words with the British who did not want the canal built for their own empirical reasons. In 1875 the British then did a complete about face and purchased a reported 44% of the shares in the canal from the Egyptian government. A blatant case of – If you can't beat them, join them.

In its first year of operations (1870) the canal averaged two ships a day. On the peaceful morning we drove south from Port Said, less than 100 years later, we saw 26 ships between the Mediterranean and the road turn-off to Ismailia – and that was before mid-day. The canal zone has not always been so tranquil.

Egyptian President Gamal Abdul Nasser nationalized the Suez Canal late in July 1956, causing a brief war between Egypt and the triad of Israel, France and Great Britain. The canal remained closed until March 1957. In the short time since then it has operated efficiently under full control of the Egyptians.

As we drove we often passed farmers irrigating their fields by the use of traditional shadoufs – a weight and balance labour-saving device that has been in use in Egypt unchanged since the time of the pharaohs.

I thought Damascus would have been a reasonable indicator of what Cairo might be like. I was wrong. There was little similarity. Cairo was the busiest city I had ever seen. Somehow Dennis navigated through the crowded streets to the Nile Hilton without incident, although I had to close my eyes a few times. We parted there at the Hilton. He shook my hand, wished me luck, tossed his keys to a concierge and strolled into the elegant hotel as if he owned it. Perhaps he did or was at least a major shareholder.

I went in search of more humble accommodation. A youth hostel was marked on my city map within walking distance. On the way I ignored the countless begging hands and the offers to, "Carry your bag, effendi, small price."

The Youth Hostel Association, which had sheltered me in towns and cities across Europe, welcomed me to Cairo. I had a clean bed in a dormitory, where my backpack would be safe and, in return, I paid a small fee. I, and all the other overnighters, agreed to do a little work each day. The custodian of the hostel assigned me the task of sweeping out the dormitory each morning. Not an exacting chore.

Cleopatra was buried with Marc Antony at an unknown location in or near Alexandria. I couldn't visit her grave, and the Pharos of Alexandria was long gone from its foundation – due to a series of earthquakes in the 14th century, however, I felt I had to travel back to the Mediterranean and see Alexandria, even if only to remind myself of the importance of the city port for Cleopatra, and for Alexander who founded the city and, in a flash of inspirational ego, named it for himself.

The train fare, third class, was cheap and the railway much less tiring than waiting in the hot sun for rides that might be many hours apart. The passengers were interesting too. Mums with kids. Mums without kids. Mums with mounds of baggage. Businessmen – not yet successful, I assumed by their clothing. Some wore local dress. A few men tried western suits. They studied me as I studied them. It helped to pass the time between the two cities.

Misr Railway Station in Alexandria was a mayhem of noisy, crowded humanity. I pushed my way through the masses and found cheap accommodation nearby. Everything I wanted to see, or experience, was within walking distance – in other words, within about two or three miles of the station. I paid to store my pack in a locker, assured by the custodian it would be there when I returned.

My mental list of places to visit included the location where the Pharos of Alexandria once stood. Close by was the 15th century Citadel of Qaitbay. West of the station was the remains of the Serapeum Temple and Pompey's Pillar. Beyond, only a short walk are the Catacombs of Kom el Shoqafa. The Catacombs became my starting point because I was thinking about Cleopatra and Marc Antony.

Legend says the catacombs were discovered by accident in 1900 when a donkey hauling a heavy cartload of stone fell into a sinkhole. At the bottom of the hole workers found an ancient temple and tombs cut into the bedrock. There is no mention in the legend of the fate of the donkey. I didn't care whether the donkey story was true or not. It was a good tale and made sense. I wanted to explore the catacombs for a different reason.

As an underground necropolis, believed to have been started for a single family in the time of the pharaohs and then expanded, it has sarcophagi containing mummified bodies – Egyptian, and urns holding cremated remains said to be Greco-Roman. As the burial place of Cleopatra and Marc Antony has never been discovered, perhaps, I told myself, they were buried in some way in the catacombs at Kom el Shoqafa. There was no way of knowing, but my fanciful version worked for me and made my hours underground in the temple and with the tombs that much more exciting.

In daylight again, I walked to the Serapeum Temple site where I wandered the ruins for an hour. They were less interesting than I had hoped but Pompey's Pillar held my attention. No one could tell me why the solid pillar was named for Pompey. It actually dates from Diocletian's time and was built in 297 A.D. At almost 100 feet in height and towering over the adjacent ruins of Serapeum, the red granite obelisk was crafted from stone believed to have been shipped down the Nile from Aswan, probably in two pieces.

Considering the weight (the pillar is estimated to be a hefty 250 tons) the barge, or barges, that transported the granite shaft must have been enormous and powered by an army of slaves at the rowing benches. What a sight that would have been.

The Mediterranean Sea laps at the rocks where the Pharos of Alexandria once beamed its warning light for mariners. One of the seven wonders of the ancient world, the lighthouse was built a couple of hundred years before the Christian era and stood for about 1,600 years. Now that's a lesson for today's architects and builders. The earthquake that toppled the lighthouse must have been localized because it had little or no effect on the pyramids only 120 miles away to the south.

Looking at the site it didn't take a genius to work out that the lighthouse, in its entirety, or in pieces, must be buried in the silt of the harbour basin or just outside on the Mediterranean seabed. I have no doubt that some day an underwater archaeological team will search the area and divers will find remnants of the Pharos.

After a few days roaming Alexandria, without finding the earthly remains of either Cleopatra or Marc Antony, I knew I was tackling a lost cause. No Cleopatra and no Marc Antony, although Alexander was very much in evidence.

The history of Egypt is overwhelming. When Alexander and his army marched across the Sinai to the Nile, Egypt's history already leaned back more than 3,000 years. Dynasty after dynasty ruled and left their marks on the people and on the land, including the first pyramids built near Saqqara. Those dynasties were followed by an era archaeologists call 'The Old Kingdom', the time of the pyramids of Gizeh, which lasted about 450 years between 2,600 B.C. and 2,150 B.C.

Then came 'The New Kingdom', beginning in about 1,600 B.C. and lasting until 1,100 B.C. Note, that period of 500 years was still 800 years or so before Alexander and an additional 250 years before Cleopatra and Marc Antony heated up the desert sands with their passion.

To put that into a different context, Egypt had a flourishing civilization of architecture and culture, and a vibrant wooden boat-building industry on the banks of the Nile, before Europeans had developed coherent patterns of speech: and long, long before the indigenous peoples of North America had built anything more substantial than skin tents, or flimsy canoes.

Alexander's army were the first known Europeans to settle in Egypt. They were Macedonians and Greeks. The Romans came a little later. The British and French added their meddling ways to the Egyptian mix in the early to middle 1800s and, in my opinion, they stayed much longer than they should have. The pyramids, however, and other monuments from antiquity, have survived all conquests and stood guard over Egypt for thousands of years and they will continue to do so, although the poor Sphinx has suffered in recent years. A local guide told me that RAF fighter jets had used the Sphinx for target practice during the Suez crisis in 1956. Insensitive cretins.

TINKER GO WALKING!

CHAPTER EIGHT

With my acceptance of the loss of Cleopatra and Marc Antony came the mental reminder that Alexander is said to have followed flocks of birds from the Mediterranean coast to an inland oasis glorified by the Oracle of Amon. In Memphis (now about 20 miles south of Cairo) Alexander had been accepted as the new ruler of Egypt. As such he became Pharaoh – son of Ra, God of the Sun. It was only fitting, therefore, that he should undertake the journey to Siwa Oasis to meet the priests of Amon.

The army marched along the Mediterranean coast for many days to reach the town of Mersa Matruh, later to be known to the Romans as Paraetonium. From there, following those birds inland, Alexander continued to the oasis. History has not recorded the impact of Alexander's army personnel on the remote, tranquil oasis in the dunes where birds followed their instincts to breed. We know the Italian and the British army soldiers made a mess of the place during WWII.

I debated on the feasibility of making the long trek. The coast road should not be a problem. I knew there were trucks and occasional buses on that road. Getting from Mersa Matruk to Siwa could prove difficult. That stretch of merciless desert, close to 180 miles, might be my downfall. I decided to ask for expert advice. I asked at the British

Consulate with negative results: "Oh, my goodness. No, you don't want to go there, old chap. It's a frightful place."

When someone tells me I don't want to go somewhere I am immediately interested in going there. It is a perversity I cannot control. I talked to an archaeologist at the museum. He was much more informative than the British Consulate but, he cautioned against the journey citing lack of transport to and from Siwa. The negativity grated on me. Alexander's men walked there. If they could do it, so could I.

Yeah! Tinker go walking!

I bought my usual sustenance of dates and bread, plus water, and set out to the west on the coast road. That was the easy part. The 180 miles to Mersa Matruh took me one day, riding in a truck. I began to sense Alexander was watching over me. I checked with various local authorities about transport for the next 180 miles roughly south-west to Siwa, near the Libyan border. The police told me I couldn't go there but neglected to tell me why. That wasn't enough for me.

"Why not?" I asked.

"You need a permit."

"Where can I get a permit?"

"Cairo!"

"I'm not going back to Cairo. Why can't you give me a permit?" I asked.

"You have to pay."

At this point I was getting a little exasperated.

"How much must I pay?"

"Ten pounds."

A flash of inspiration made me ask, "How much does the permit cost in Cairo?"

Answer, spoken with a straight face. "In Cairo the permit is free."

"So, there is no fee. Correct? It's just *baksheesh* for you?"

The police officer gave a weak smile and a slight shrug of his shoulders. "It is the custom," he replied.

Ignoring the custom, I muttered an Anglo-Saxon oath, picked up my backpack and walked out. I spent the next day and a half at the road junction where a marked track pointed to Siwa Oasis. Nothing stirred in that direction. I was almost ready to give up my quest when an Egyptian Army truck rumbled along the road from Mersa. I could see it was a four-wheel-drive and that it had sand ladders on one, possibly both sides.

It could only be going to Siwa. I stood up and stuck my thumb out, a hopeless gesture but, I felt, necessary. The truck stopped. The driver asked me where I was going. I told him Siwa. He asked if I had a permit.

"No. No permit," I admitted.

"Okay. I can give. Cost five pounds."

I laughed. The situation was so perfect. So obvious. *Baksheesh*, the wonder that makes a large part of the world go around. I handed him five one-pound notes. He pocketed three and gave the other two to his companion, another soldier. He jerked his thumb towards the back of the truck as I asked, "What about my permit?"

"No is necessary," he replied. "You have paid."

I laughed at him and jumped into the empty back of the truck. Two long days later, after a gruelling ride over an abominable track across the Libyan Plateau and a long stretch of featureless desert, we arrived at Siwa, about 80 feet below sea level. On the way we had passed a sad selection of burned out vehicles that had not survived the war.

"How long are you staying here?" I asked.

"We leave two days, maybe. Take soldiers back to base near Cairo," the driver answered. "You want ride?"

"How much?" I asked with a certain resignation.

"Five pounds," he answered, "for the permit."

I was quite sure the army truck was my only option for the immediate future, so I agreed. I didn't have any time to waste. There was much for me to see in, perhaps, only one full day.

Siwa Oasis was worth the difficult journey, and the ten pounds it cost me to get there and back. Majestic sand dunes stood in silent menace to guard the oasis. I knew they were dangerous. That didn't stop me from wanting to explore them. First though, I had to find the Oracle of Amon, in ruins now but still a glorious reminder of Alexander to me. That site kept me busy for the few hours of daylight left to me, and a few more in the morning. I have a passion for scrambling around in dead cities with strong historic tales behind them. In my explorations my imagination runs wild. I can cast myself back to be among the populace when Alexander arrived. In doing so I create my own memories of historical events and disturb no one, not even the long dead, except a few lizards, some scorpions, and an occasional sleeping dog.

With time running out, I hastened to Shali Ghadi where a fortress said to be nearly 800 years old overlooks the rubble of long-gone habitations. I was in my element.

Despite my unseemly rush, I enjoyed Siwa. The oasis is a lovely garden of olive and palm trees fed by underground springs that bubble into sheltered ponds. I stripped to my underwear and went for a swim, and a wash, finding the water cool and clear.

The big sand dunes nearby looked inviting, as I said. I would love to have had the time to ride through them on a camel, or a donkey. The army put paid to that idea. I met the truck driver in the early evening at a tea shop and he told me to meet him there at sunrise, ready to travel. With his agreement, I slept in the back of the truck that night so as not to get left behind.

Alexander and his men returned to Memphis directly from Siwa by skirting the south side of the Qattara Depression. I didn't have that option, although I would have preferred it. I had no choice but to return by way of Mersa Matruh with the soldiers.

Ten soldiers joined me on board the truck next morning. They were tough and they were tired, having been on a training exercise in the dunes for a few weeks. Their interest in me waned soon after we climbed out of the depression and began the sand crossing. When we broke down the first time they were all asleep. A fierce banging and rattling and thumping on the underside of the truck woke them and told me we had serious trouble.

We all bailed out as the truck ground to a halt in soft sand. Suddenly everyone was talking at once. I crawled under the truck where the driver sat like a statue with an adjustable wrench in one hand. I could see why. We really were in trouble. The drive shaft from the transfer housing to the rear differential had pulled out and windmilled a deadly circuit to sever the brake lines. I took the wrench from him as he crabbed out to speak to the soldiers, which he did in rapid-fire dialect.

While he was thus engaged, I removed the drive shaft from the transfer housing. Although I was sure of what I would see, I unbolted the differential cover. A handful of oily gear teeth dropped onto the sand. There was nothing I could do there. I replaced the cover, picked up the now sand-covered teeth and dragged the drive shaft behind me into the sun.

The soldiers were busy. They had removed the sand ladders and were digging tracks to get all four wheels out of the sand. I pulled the driver aside and showed him the teeth, explaining where they were from. He groaned.

"We can still run on front wheel drive, if we can get the truck out of the sand," I told him.

He gave me a quizzical look.

"Put the gearbox in four-wheel-drive. Only the front wheels will be turned. The rear wheels will follow like any other truck," I explained. It would be noisy, with those teeth missing but it would work – for a while.

With much revving of the engine by the driver and with the rest of us pushing with all our might, we got the truck out of the sand trap and up onto the sand-ladders. These items are flat sheets of metal with holes cut in them. We had enough in front of the wheels for three vehicle lengths. I hoped it would be enough. Once the truck started moving the driver would not dare to stop for a second, otherwise he would lose all traction.

I walked ahead with two of the soldiers, checking the ground for stability. Between us we marked out the most driveable route to a few hundred yards of sharp gravel that felt stable enough to take the truck. A soldier waved the truck on. With all of us pushing, the driver managed to reach the gravel and comparative safety. By the time we had retrieved the sand-ladders and restowed them, the day was almost over. We decided to sleep there beside the truck that night.

That evening I was treated to a filling meal of rice and goat's meat, with hot tea and a handful of dates each. We sat around the fire for a couple of hours while some of the soldiers sang their favourite songs. I assumed they were all love songs because the only word I recall is *habibi*, which means something to do with love.

We took two and a half more days of slow, careful driving to reach Mersa Matruh. There the truck shuddered to a halt. The strain on the front differential had been too much. However, we were the army and, as I soon learned, that gave us extra powers. We were soon towed to a garage and set on a primitive hoist. There, three of us, the truck driver, a mechanic from the garage, and I, changed the differentials. Working together, with a few breaks for tea, it took us all day.

While we were working on the truck a uniformed policeman wandered over to the garage. I recognized him immediately, and I'm convinced he recognized me. He was the one who had tried to extort ten pounds from me for a non-existent permit. He glared at me for a long time before turning away. I suspect the presence of the soldiers had something to do with his decision.

Six days after we left Siwa Oasis I dropped off the back of the truck in Cairo's northern suburbs. Waving farewell to my military companions, I walked in the direction of the distant hostel.

Feeling lost and lonely early one evening soon after the Siwa expedition, I walked to All Saints Cathedral, wishing to sit quietly in a pew and think, perhaps to pray. The church was locked. I went to the vicarage next door. An English lady came to the door. When she heard what I wanted she said, "We're closed. Come back on Sunday. Service times are posted beside the main door."

Without meaning to be impolite, I said, "That's not exactly Christian, is it? I understood God's house was always open."

The lady frowned at me, raised her eyebrows and made some half-hearted excuse about difficult times. As I turned away, before she shut her door, I said, "I'll go to a mosque instead. Mohammed's house is always open for prayer."

And I did. I left my sandals at the door with a handful of others, tip-toed in and sat cross-legged on a carpet with my elbows on my knees and my chin cupped in my hands. I stayed there, thinking and drawing comfort from the quiet until I heard the Imam calling the faithful to prayer from the minaret. I left as quietly as I had arrived. I felt much better.

CHAPTER NINE

Brandy scorched into my life in Cairo and left me exhausted and exhilarated in Luxor a month later. I was in Egypt to learn. She was there to play. Brandy's father worked at the American Embassy. She was visiting Mummy and Daddy for a few weeks before going home to attend college, either Vassar or Bryn Mawr. I've forgotten which one.

She came from a small, Victorian-looking town, she explained, on the north coast of California. "Eureka," she told me. A town, not an exclamation of pleasant surprise. Despite her small-town start, Brandy was a worldly young vixen who spoke in short, fast sentences. She was hot-blooded and fearless. It was a volatile combination on sultry nights in the Nile valley. She reminded me of Molly, except that I knew Molly was married by then and had a daughter. She reminded me of Molly because they both liked to instruct.

"Touch me here! Kiss me here."

I was still learning.

We met at the Sphinx. I was alone, examining the enormous scarred limestone figure, with a pen and notepad in my hand, when a group of American voices approached.

"Hi," a bright greeting interrupted my thoughts. "Are you an archaeologist?"

I turned and was captivated by a tall, slim figure in a floral dress, with a white shawl over her head and sandals on her otherwise bare feet.

"No," I answered. "I'm just a traveller enjoying history."

She smiled, showing a lot of perfect teeth, and stuck out a hand. "I'm Brandy," she said. "And you are?"

"Hello. Everyone calls me Tinker."

A spark crackled between our hands as her long fingers touched mine. We laughed in unison.

"Why are you called Tinker?" she asked with a big smile.

"Because I'm good with my hands. Good at fixing things. I like tinkering with anything mechanical."

"Okay, where are you staying, Tinker?" Brandy asked.

"I'm at the Youth Hostel in Cairo."

"I know where that is. I have to go now. Would you like to come back here tomorrow with me? There's so much more to look at," she was bold and a far cry from the shyness of other girls I had known, except Molly. I liked her.

"Okay," I said. "I'll meet you at the hostel about nine a.m."

We met as planned. Brandy was on time and shook my hand without setting off sparks. We took a local bus to Gizeh – a new experience for her. I don't think she had ever been on a public bus. She wore a different cream-coloured dress and a matching shawl draped around her shoulders instead of over her head. She had straight dark hair cut short to reveal her slender neck. Brandy had hazel eyes.

That morning I learned her real name was Brenda, but she preferred Brandy. I thought she was pretty. Side by side during one long day we went back to the Sphinx, which represents Pharoah Khafre, and we explored the bases of the three pyramids, burial monuments for Khafre, Kufu (Cheops), and Menkaure, all built between 2,550 and 2,490 B.C.

As the sun merged with the Sahara in a blaze of red and gold at the end of the day, we took a bus back to the centre of Cairo. Brandy fell asleep with my arm around her and her head on my shoulder. I left her at her home after agreeing to meet again the next day to climb the great pyramid.

The great pyramid of Cheops is the largest of the three at Gizeh. We held hands in the dark and claustrophobic interior tunnel leading to the cavernous yet disappointingly empty funeral chamber. It was cool and dark in there, with nothing to see. Out in the hot desert air again in the early afternoon, we climbed the large limestone blocks to the summit.

The stones on the route we took up one corner were worn smooth from the passage of feet over time. On either side the identical stones showed rough edges, untouched for centuries by anything other than the desert winds and scouring sand. The climb was hard, but it was worth it. We had the summit to ourselves.

On top, where the fertile Nile valley lay below us to the east, and the Sahara stretched its sands north, south and west, we stayed overnight and, to my surprise – and pleasure, made love under the stars. Brandy claimed we had done it on Ptolemy's rooftop. She didn't care about historical accuracy. As far as she was concerned all pharaohs were Ptolemys, even Cheops. The fact that Ptolemy was a Greek Macedonian from Alexander's time and Cheops had ruled Egypt over 2,000 years before him failed to change her thinking.

Looking up at the massive pyramid from ground level later I considered us lucky not to have rolled off the flat summit rocks. I could almost see the newspaper headlines: "Two naked tourists killed while rolling down from the top of Egyptian pyramid."

Concerned that Brandy's parents would be worried about her, I offered to explain to them that I had been looking after her.

"You don't need to do that. I doubt that either of them would have noticed anyway. My father is always late home – sometimes I don't see him for days. And Mother lives her own social whirl of a life. I do as I wish."

For the next two weeks we were inseparable. Brandy's self-centred parents gave her the freedom. Already her lover, I became her bodyguard as well. Cairo was our playground. We roamed the bazaars by day and by night. We spent a few days among the thousands of artefacts in the Egyptian Museum of Antiquities. One day we took to the river, drifting with the current in a rented felucca, a traditional Nile sailing boat, from Gizeh to the northern suburbs. Most evenings Brandy had drinks or dinner with embassy friends, occasionally with her parents, while I dined in a tea shop in the bazaar.

With approval from her parents who, she said, were too busy to get away from Cairo for a week, we took an overnight train to Luxor. Travelling in a second-class sleeper: Brandy piled our bags on the lower bunk – my backpack and her two suitcases. We climbed up and shared the narrow top bunk: I almost broke my back when we fell out together while in the throes of passion. I groaned, "Ooof!" as I hit the floor of the swaying carriage. Brandy landed on top of me and complained with a most unladylike, "Fuck!" as we fell apart. Then we started to laugh.

I don't know what she was thinking, but I had a vision of that pyramid in my mind.

"We should have used my bunk," she spluttered when we got some air back into our lungs. We finished our tryst on the floor, helped, no doubt by the vibration of the train. My back was sore for many days after that.

Luxor is on the east bank of the Nile. The Valley of the Kings was on the west side. We checked in to a modest hostelry, left our bags in our room, and went exploring.

"Where shall we go first?" asked Brandy.

"The Temple of Karnak," I answered. "It's not far. We can take a gharry for a few cents."

Seated in the horse-drawn carriage like minor royalty, we waved to people on the streets, some of whom waved back. The horse behaved itself, not stopping once to embarrass us. We left it at the entrance to Karnak. While I paid the driver, Brandy held the horse's face in both hands and kissed it on the nose. Seeing my look, she explained with a shrug of her shoulders and a smile, "I like horses."

The site of Karnak is enormous. The temple alone is ranked as the largest religious building ever created. It covers a footprint of 200 acres and was a place of pilgrimage. Dedicated to the gods Amun, Mut and

Khonsu, Karnak was, without doubt, one of the most magnificent buildings the world has ever seen. In its original state, decorated with scenes and hieroglyphics in bright colours, its majesty would have convinced even the most devout disbelievers of the presence of deities. Brandy and I saw Karnak as a collection of broken walls, pillars and statuary. We also saw it for what it once was, a place of beauty which could only be the home of Gods and Goddesses. Karnak inspired Brandy to suggest a night on the temple-town.

"Let's come back after sunset and stay here?" she begged. "We can bring your sleeping bag and stay awake watching for ghosts of ancient Egypt."

We did spend the night at Karnak. We saw no ghosts but, in the early hours of the morning, while Brandy slept with her head on my shoulder, I sensed I could hear the soft sandals of the long-dead pharaohs gliding through the temple with their priests. It was an eerie feeling, but it gave me a comfort I could never describe.

With another couple, both from Scotland, we hired a boatman to take us across the Nile to the wonders on the other side. Close to the west bank our boatman pointed out the snout and eyes of a huge crocodile basking in the shallows.

"Crocodile much bloody bad," the boatman told us. We agreed. 'Crocodiles were much bloody bad.'

The Scottish couple, a decade or more older than us, said they were working in Cairo training young Egyptians to be bankers. The visit to Luxor was to be a final look at Egypt as it once was before returning home to Scotland.

We rented a 1926 Ford taxi with open sides and a torn canvas roof for the day. The driver, old enough to have been my great-grandfather, was not the safest custodian of a motorized vehicle I had ever encountered. He had a forked stick to hold the gearstick in place and, as I sat in front with him, he said I should move the stick each time he changed gears, which he did more often than necessary. The lack of traffic on the sand-covered gravel road gave us a sense of security that the driver did his best to dispel. Despite the best efforts of our genial daredevil behind the wheel, we survived a long day of roaming the lands of the dead on the west bank of the Nile.

The gigantic, though time-worn and tired Colossi of Memnon greeted us as we rattled towards the Valley of the Queens. Those two seated statues, both 60 feet tall, have kept watch over their part of the Nile Valley opposite Luxor for over 3,300 years. Now faceless representations of Pharaoh Amenhotep III, they greet the rising sun where they were placed in front of Amenhotep's mortuary temple, although little of that building exists, except rubble.

"But why are they called Memnon if they are meant to be Amenhotep?" asked Brandy.

"The simple answer to that question is that Greeks from eons ago mistakenly associated Memnon, a Greek, with Amenhotep, the Egyptian, and no one has seen fit to correct that error."

With a grinding of gears and a little dexterity with the forked stick holding the gear lever in place, we moved on to the Valley of the Queens tucked away in a gorge. An estimated 90 or more tombs of wives and daughters of a succession of pharaohs reside in Ta-set-Neferu, the correct name for the burial site. We were told the name means 'Place of Beauty.'

There was so much to discover in ancient Thebes. I could have stayed for months and not studied everything. However, my time was limited by Brandy's need to return to Cairo at the end of the week. We left the dead royal ladies to their eternal rest and drove the short distance to the most impressive building in ancient Thebes.

Queen Hatshepsut's temple tomb at Deir el Bahari is 3,000 years old yet it still stands as a glorious example of ancient Egyptian architecture. The other three went ahead of me. I held back to spend more time just admiring the temple set against a rugged sandstone cliff. I needed to take in the complete vista before walking the avenue to the straight, inclined road leading from the ground to the next levels.

Others are buried in the mortuary temples at Deir el Bahari, but none so special as the formidable Hatshepsut who reigned over Egypt for 22 years.

I walked up the ramp to the second level and slowly approached the second ramp where my companions waited. Brandy pointed to the next level.

"Look at the figures up there," she said. "There are eight of them, and a long line of columns."

"Those figures, or colossi, are giant images of Hatshepsut," I explained. "Let's go and meet her."

Deir el Bahari

The colossi of Hatshepsut stand taller than three men of my size. They appear to be either guardians or welcoming figures to the inner temple.

"Did you know Hatshepsut ruled Egypt fourteen centuries before Cleopatra?" I was reading from my poorly printed map guide to Deir el Bahari. "She was originally buried somewhere over there, in or near the Valley of the Kings." I pointed beyond the sandstone cliffs.

While the others studied the columns and effigies of Hatshepsut, I continued reading. "Below us, on the side of the ramp, is the story of Hatshepsut's expedition to the Land of Punt in 1493 BC. Hatshepsut's mother, Hathor, is said to have been born in Punt. Unlike most Egyptian expeditions of the pharaonic era, the emissaries were sent there to trade, not as warriors bent on conquest."

"Where was Punt?" Brandy asked. "I've never heard of it."

I explained that, although the exact location and boundaries of Punt have never been established, scholars argue that it was in an area that now encompasses Eritrea, Djibouti, a small part of Saudi Arabia and Yemen, and a slice of northern Somalia.

I would have preferred to explore the Valley of the Kings by myself, or just with Brandy. With our two other companions that proved almost impossible, although I did escape a few times to view parts of the necropolis alone and in silence. I sat on the cold sandstone floor of the funerary chamber which once held the mortal remains of the young pharaoh Tutankhamun and I leaned back against the wall. I closed my eyes and tried to imagine the scene that day in 1323 BC when the young pharaoh was laid to rest.

> Muted drums and wailing reed flutes announce the approach of a grand funeral barge bearing a human-shaped coffin decorated in blue and gold and bearing the likeness of the recently deceased Pharaoh Tutankhamun. The ranks of rowers dip their oars in unison, responding to the slow, mournful tempo created by the drummer standing on the stern. At a landing stage on the west bank of the Nile a host of dignitaries, most of them priests, await the sad arrival. Hundreds of bare-chested slaves, clad only in sandals and short kilts, stand beyond in silence. As the coffin containing the mummified remains is taken ashore the priests lead the way on foot towards the distant tomb waiting in the Valley of the Kings. The slaves follow bearing the dead pharaoh. Hours later, at the entrance to the richly decorated tomb, a huge Nubian slave enters first, carrying a fiery beacon of oiled reeds. He lights a series of smaller beacons set into the walls. The priests come in two by two and stand in silent ranks on either side of a large stone sarcophagus. The blue and gold likeness of Tutankhamun is lowered into the cold stone and a massive lid requiring ten slaves to lift is placed over the top.

I open my eyes and wonder how so many men, plus the pharaoh's body, had squeezed into the chamber. Brandy's voice called me from my reverie. She was outside in the hot sun, shading herself under a large yellow umbrella.

"Where did you find that monstrosity?" I asked.

"I've been looking for you," she countered.

"I was in there," I pointed. "In Tutankhamun's burial chamber. It's much cooler in there. I was thinking about the day the pharaoh was buried and imagining I was there."

Brandy grinned up at me. "Oh, Tinker," she said. "You are such a romantic, and a dreamer."

Brandy was right, of course. I was a romantic and I was a dreamer.

Discovered and uncovered by British archaeologist Howard Carter's team in late 1922, Tutakhamun's burial place is the focal point of the Valley of the Kings. It is also said to be cursed. Anyone who defiles the boy pharaoh's tomb will suffer a terrible death. Does that include me? I wondered. All I did was sit there and dream. I found the burial chamber to be a peaceful place. No defiling there.

In retrospect, having wandered through the valleys of the kings and queens, and many of the other necropolises on the west bank, Queen Hatshepsut's temple at Deir el Bahari is the one that stands out in my mind.

Brandy and I parted company in Luxor. She went back to Cairo by train with the Scottish couple. I was tempted to go with her. Commonsense dictated that I stay. Our passion was at its height, but we were not in love, except – maybe – a little bit. Our lives had crossed by serendipity, and we had had fun together. In reality, though, our futures were destined for far different passages. We left each other at the railway station with only a few tears and no promises. Brandy leaned out of the carriage window and kissed me.

"Look after yourself, my Tinker," she whispered.

"Yeah. You too, my Brandy," I said.

We both knew, if our paths should cross in the future, we would greet each other warmly, maybe with a few sparks, but again with no promises. A whistle blew as Brandy called out, "Hey. You never told me your real name?"

I laughed and shouted back, "You can call me Tinker."

I heard her laughter as she pulled back from the window and took her seat. Fate decreed that we should never meet again.

I had my pack with me. As Brandy's train chuffed out of the station, I hauled the pack onto my back and set off on foot in the opposite direction to look for a ride south. We had stayed awake all night, holding each other; touching each other. I was tired, but not too tired for a walk. I was suddenly lonely, too. Very lonely, but I had dreams to fulfill.

TINKER GO WALKING!

CHAPTER TEN

"You'll never get to Khartoum going overland," an expert in Cairo had told me. "There are no roads," he added.

I asked again at Luxor and received a similar answer. Undeterred, I hitched a ride to Aswan a few hours after Brandy left. At Aswan I asked again.

"Are there any trucks, or cars going to Wadi Halfa?"

The answer was the same. I would like to have gone by river steamer the length of Lake Nasser to Wadi Halfa and then by train to Khartoum, but I couldn't afford the fares. I needed a job for a few weeks. There was construction work beginning all around me. Work was starting on a new dam to change the water levels on the upper Nile.

"No foreigners," I was told. "We can only employ Egyptian workers."

I tried to get a temporary job with a local trucking company, either driving or repairing vehicles. No jobs for itinerant foreigners there either – except – "If you could get that truck started, I could pay you a little."

The truck, an old Dodge Power Wagon, had seen much better days and more than a few accidents. I raised the engine cover and it fell off, the bolts holding the hinges having been removed. The truck's engine was a mess. Rodents had nested in there. The battery was long dead and unlikely to respond to resuscitation.

I spent the first day getting rid of the nests, checking the wiring, and cleaning every part of the engine I could reach while I charged a spare battery I found in the garage. The second day I removed, cleaned and replaced all the sparkplugs and the filters. I changed the oil and I drained the little fuel left in the tank into a large can. With a bent wire brush, I scraped the inside of the fuel tank as much as possible to remove any foreign particles before pouring the fuel back in through a filtered funnel.

On the third day I used a hand-crank to persuade the engine to start, which it did with many bronchial coughs and asthmatic wheezes. When it fired into full voice, delivering an impressive cloud of black smoke, the owner rushed out of his office to stand and stare in amazement. I refitted the engine hood and took the Dodge for a short drive. When I brought it back the owner climbed into the cab with me. He rudely pushed me over to the passenger side, turned off the engine and then, with a muttered prayer to Allah, he turned the ignition on and pumped the gas pedal a little. The truck started. The owner beamed with pleasure. In his office he handed me one Egyptian pound as pay. I shook my head.

"Not enough," I argued. "I worked almost three days on that piece of junk. You owe me at least five pounds."

The ensuing argument soon collected a small crowd of amused citizens of Aswan. Some said one pound was enough for a foreigner. A few others knew I was being cheated and said so. The argument swayed back and forth. It was not going in my favour, I decided, so I went out to the truck, removed the keys and threatened to throw them in the Nile near some crocodiles. The truck's owner almost had a fit. We returned to bargaining, the keys jingling in my hand. The audience crowded as close as they could, determined not to miss a single nuance of the entertaining drama.

Coin by coin the offered payment rose through the pounds to about four and a half. The owner gave the impression he was dying because of my avarice. To allow him to save face in front of the crowd, I agreed to take the four and a half pounds – about thirteen American dollars. We shook hands on the deal. He paid me the money. I gave him his keys. The audience applauded and drifted away in twos and threes. The show was over.

I had more money, though still not enough for the ferry and food. No trucks were heading south into the desert.

"No one goes that way," I was told again.

I considered returning to Cairo, spending a few more days with Brandy, and then taking the coast road through Libya and Tunisia to Algeria before crossing the Sahara to Timbuktu. It was a reasonable alternative, but I did want to see Khartoum. I decided to be patient for a week. Maybe something would happen to help me travel south.

It was a sensible decision. My luck was about to change. Between the Corniche el Nil and the riverbank, where feluccas gathered opposite Elephantine Island, just north of the first cataract of the Nile, I met a German in trouble. More important, for me, he was a German in mechanical trouble. Standing on a jack with one rear wheel in the air, was a Volkswagen Kubelwagon from WWII.

Painted in a colour to match the desert sand and with extra fuel cans in racks on either side, it was a 4-door, 4-seat open car, with a canvas top folded back. It was a lightweight vehicle with an air-cooled engine that could cross deserts without the need of water and without burning much fuel. I moved in for a close look.

"What's the problem?" I asked of a young man sitting on a box nearby. I knew he had to be with the car. He had blond hair and blue eyes and was dressed in khaki shorts and shirt.

His few words of English were about as useful as my few words of German. We resorted to sign language. That was more effective. He showed me the damage. A shock absorber had been removed.

"Is kaput," he said.

He managed to tell me his friend had gone looking for a mechanic to help them.

"He sprachen English," he said. Useful information. I sat and waited. I learned his name was Willi and his friend was Jurgen. I understood they were going to Khartoum and that Jurgen owned the car. I sensed I was looking at my ride to the south.

When Jurgen returned he looked dejected. He had two pieces of metal in his hands – two halves with no value. Willi introduced us. Jurgen was preoccupied, explaining the mechanic at the garage had said he must order the part from Germany. I had a better idea.

Leaving my pack in the car under Willi's care, Jurgen and I went in search of a welder. We found one near the bus station. He studied the two pieces, fitted them together, took them apart, shook his head and said, "Not possible."

I asked him how much he would charge to rent his welding equipment for an hour. We haggled for a few minutes and agreed on one

Egyptian pound, which Jurgen paid. I found a slender piece of iron about the size of my index finger on the welder's workbench and tacked it to the largest part of the broken shock. When it cooled, I added the smaller part, fitting the break together and tacking that to the finger of iron. Then I welded the iron rod to the shock and put a seam right around the break. It wasn't pretty but I knew that part would never break again. Jurgen and Willi were delighted with my efforts and wanted to pay me. I shook my head.

"I want to go with you to Khartoum," I said. "If you break down in the desert you'll be in real trouble. If I'm with you, I can probably fix it."

They went for a walk to talk over this new development. I waited with the Kubelwagon, now standing on all four wheels and looking ready for action. When they came back Jurgen asked, "Can you drive?"

"Yes," I grinned at him. "If you want me to, I'll drive all the way to Khartoum." And so I did. They were planning to drive the desert route to Wadi Halfa and then, if conditions were favourable, continue to Khartoum and beyond. Their ultimate goal was Dar Es Salaam on the Indian Ocean coast of Tanganyika (now Tanzania).

Loaded with enough gasoline and drinking water to take us across the desert west of Lake Nasser, we set off on the almost 200 miles drive to Wadi Halfa. By the end of the first day Jurgen was suffering from the runs. We stopped early to make him as comfortable as possible. Willi looked after him while I checked the car. In the morning Jurgen was worse.

"We have to get him to Wadi Halfa as quickly as possible," I said. That meant no possibility of stopping for a few days at Abu Simbel and exploring the monumental temple of Rameses II. My disappointment was mitigated by my concern for Jurgen. We reached the southern finger of Lake Nasser early in the afternoon. Willi took Jurgen the last few miles to Wadi Halfa in a felucca. I waited for another day before a makeshift barge arrived to carry me and the car across to the 'road' on the east side. From there I drove to Wadi Halfa where I found my travelling companions staying with a German engineer in a pleasant villa overlooking the lake.

A local doctor had diagnosed Jurgen with dysentery and recommended he stay in the comfort of the villa in Wadi Halfa for at least a week to recover. Tempted though I was to try to continue alone to Khartoum, I chose to stay with my new friends in case Willi needed help. It was the right choice to make. While Jurgen rested, Willi and I roamed Wadi Halfa and its surroundings. Of particular interest to me was the small port on the east side of the southern tip of Lake Nasser. The fishermen and the ferrymen were hard workers, but always ready for a chat about their professions. I spent many happy hours in their company and, I believe, Willi benefited from the experience as well.

The next stage of our journey would take us across the desert to cut out the great bend of the Nile. That meant we would not see the deep gorges and the other cataracts that have always restricted travel on the mighty Nile River. I added the river route to my mental list of future adventures.

The First Cataract of the Nile at Aswan

Nine days after we arrived we left Wadi Halfa. With a refuelling stop in Atbara, we covered the 500 miles to Khartoum in three long, hot days.

My German friends only stayed in Sudan's capital city for a few days to get travel permits for the next stage of their journey to Kenya and Tanganyika. They invited me to join them for drinks with some guys from the embassy, which I did. Not surprising, the conversations were all in German, so I missed most of the information. However, I did make the acquaintance of one who addressed me in English, having heard how

I repaired Jurgen's car, he suggested he might know of a couple of mechanical jobs for me to look at.

Khartoum was not as interesting as I had expected, so I was tempted to go with my new friends. The pygmy people of the Ituri forest in the Congo beckoned – perhaps a real-life Lilliput. Equally, the tall Masai of Kenya could be my introduction to Brobdingnag. The decision was a difficult one, but Timbuktu called in a more powerful voice. I would stay in Khartoum a while longer and then, soon, I would follow the setting sun.

I hoped Jurgen and Willi would make it south through the Sudan, across Ethiopia, and then the final haul via the Tukhana land of northern Kenya to Nairobi and on to Dar Es Salaam without encountering mechanical problems. Before they left I checked the engine and brakes on the car and pronounced it safe. They departed with hearty handshakes and big smiles. Ahead of them stretched another 1,700 miles of rough tracks and roads.

I worked for a while in Khartoum, doing odd jobs on cars and motorbikes for the various foreign legations, thanks to my contact at the German embassy, and for private citizens from Europe. I insisted on payment in dollars, American, of course. Gradually, dollar by dollar, I rebuilt my finances to allow me to continue.

A highlight of my stay in Khartoum was the opportunity of studying in the library at the University of Khartoum. Sudan has a history of being invaded from the north and from the south. I wanted to know more. The university gave me that opportunity in quiet surroundings. The library opened my mind with its wonderful collection of tomes waiting to be studied by curious minds.

Khartoum is a relatively new city, compared to the extreme age of others along the Nile. It was established in 1821 as an outpost of the Egyptian army. I never understood why the Egyptian leaders felt they needed a military garrison so far south until I studied the turbulent history of Sudan in the 16th to 19th centuries.

Sudan has seen more than its share of insurgents, some well-meaning, others more aggressive. The land we know as Sudan is vast, covering almost one million square miles. It borders Egypt, Libya, Tchad, Ethiopia, Central African Republic, Congo, Uganda and Kenya. Within those boundaries lived a diversity of tribes, each with its own land. The northern half of the country was populated by the Nubians – the blackest people I have seen. They sensibly lived close to the banks of the Nile in relative peace until the Egyptians began to push up river past the now notorious cataracts around 2,600 BCE.

They hadn't been invited but the Egyptians planned on staying. To their credit, they and the Nubians lived in reasonable harmony in a kingdom they called Kush until about 350 A.D.

Christian missionaries arrived and began converting the Kushites in the 6th century. They were followed and ousted by Muslims a few centuries later. The Funj dynasty swept into the Sudan in the 1500s. They adopted Islam and ruled, with a few hiccups, for the next couple of hundred years. Meanwhile, in the south, where the desert gives way to savannah, forests and expansive swamplands, tribal Africa held sway. Mostly animists, and a few Christians, the dominant peoples were the Dinka. I'm not sure where the Dinka came from but they, like the others, had not been invited.

Sometime in the 16th century a nomadic tribe wandered north from near the equator into Nubia. Known as the Funj, the nomads adopted Islam and, with the Nubians, created the Kingdom of Sennar. In the west the Sultan of Darfur was increasing his power. Sudan, because of the Nile running through it from south to North, was becoming known by outsiders and gaining in popularity.

In 1821 the Egyptians under Ibrahim Pasha (an adopted son of the Ottoman ruler) took military possession of the site that became Khartoum at the confluence of the Blue and White Niles. In less than two decades Khartoum grew to a population of an estimated 30,000 people, most of whom lived in mud huts. The Ottoman Empire, of which Egypt was then a part, did not look after the Sudanese. Not surprising then that a figure rose up who was prepared, he claimed, to lead them and improve their lot. His name was Sheikh Khalifa Mohammed Ahmad.

A charismatic religious leader born in Dongola, on the Nile in northern Sudan, in 1844, Mohammed Ahmad was the son of a boat-builder. Highly intelligent, a devout Muslim and something of a mystic, his charisma attracted like-minded followers with ease. By the early 1880s he had a vast army willing to join him in overthrowing the Ottomans, the Egyptians, and the British who advised them. He referred to himself as The Mahdi, the 'Guided One.'

General Gordon was a handsome British army officer in the Royal Engineers. He was medium height and known for serving his country well – often far from home and under extreme conditions. He became famous for his heroic defense of Shanghai during the Taiping rebellion.

From that he was awarded the unofficial title of 'Chinese' Gordon by the British public. In 1873 he accepted employment from the Khedive of Egypt as Governor of the Sudan province of Equatoria. During his tenure, which lasted from April 1874 to December 1876, he mapped the upper Nile all the way to Uganda, and he was successful in reducing the slave trade.

Gordon, who could have been a double for Hollywood actor Paul Newman, returned to the Sudan in 1884 on a British government mission to evacuate Egyptian troops from Khartoum, which was being threatened by the Mahdi's army. The British chose not to defend Sudan and moved most of their soldiers down the Nile from Khartoum to Egypt. Only General Charles Gordon remained behind with a weak garrison of a few British and native troops in a futile bid to keep Khartoum out of the Mahdi's hands.

The city was under siege for a year, until a force of 50,000 Mahdist warriors attacked late at night in January 1885. Weakened by hunger and vastly outnumbered, Gordon's men were unable to resist the onslaught. Every man died, including the 51-year-old General Gordon, plus an estimated 4,000 citizens. Muhammed Ahmad, the Mahdi, died some six months later of natural causes, although he was only in his middle years.

The British and Egyptian consortium left Sudan alone for a decade and then Major General Sir Herbert Kitchener was sent to retaliate against the Mahdi's forces in 1898. He took with him 8,200 British army regulars, plus more than 17,000 combined Egyptian and Sudanese soldiers. Leaving Egypt in the early spring he moved his army south. To bolster a fleet of boats on the Nile, he had a single-track railway built running parallel to the river.

While Kitchener's force advanced Khalifa Abdullah al-Taashi, the successor to the Mahdi, led his army north to meet the British near Atbara. Kitchener forestalled him and pushed him back. Khalifa retreated to Omdurman. Kitchener followed, camping his force on the banks of the Nile a short distance north of Omdurman. While his scouts spied on the enemy, he sent gunboats upstream to blast Omdurman with shells.

Kitchener had chosen his battle site with care by taking the fight close to the Nile north of Omdurman. His army consisted of a reasonably large contingent of trained warriors armed with rifles and cannon. Except that facing them across the blistering plain was a desert-hardened force of 50,000 untrained fighting men of the Khalifa's religious army. They came out of Omdurman's back streets on two sides in a bid to encircle Kitchener's men and their formidable weapons.

The superior numbers of the Khalifa's army were no match for Kitchener's modern armaments. The British prevailed on September 2nd in a day-long battle that became a massacre.

Having won the battle for Omdurman, Kitchener rearranged Khartoum. He laid out with streets on the diagonal leading to roundabouts, or traffic circles, like the spokes of a wheel. Such a layout being easier to defend than a rectangular grid system, although one 'expert' declared that Kitchener had laid the city streets as a copy of the Union Jack.

Britain ruled Sudan jointly with Egypt until the end of 1955. On the first day of January 1956 Sudan became an independent republic.

For few decades in the mid-1800s under Egyptian rule Khartoum was a thriving slave trading city. An estimated 40,000 undeserving humans, men, women and children, were sold here each year. The slaves were African, a high proportion of the traders were Arabs. In the years between Gordon and Kitchener, slavery again reared its ugly head.

Arab nations made extensive inroads into rural Africa in the 17th and 18th centuries, not as colonizers, rather as slave traders and, earlier, to spread the words of the Prophet Mohammed. European nations took the business to a different level. They carved up Africa to suit themselves without any consideration for tribal boundaries.

The colonization of Africa (and other continents) by European powers has disturbed me for a long time. My 20-year-old brain could never understand the rationale of a man dressed in completely inappropriate clothing for the climate planting a Union Jack, for example, on a foreign shore and announcing to the watching inhabitants of that country – "I claim this land in the name of His/Her Majesty, blah, blah, blah."

There's something rather pathetic about such a scene. I imagine a small sailing ship offshore with, perhaps, 60 to 100 men on board. They are thousands of nautical miles from home and the land they have the arrogance to claim is populated by, say, 100,000 people. The very idea makes me want to rewrite history. I want to be standing on one of those foreign shores with my fellow citizens and watch one of those pompous, self-important emissaries plant his flag and make his outrageous statement, just so I and my fellow citizens can say to him in loud, clear voices, "No. We don't want you here. Get back in your boat. Go home. This is our country." There is a much more concise version of that statement which can be expressed with greater emphasis in two often-joined Anglo-Saxon words.

Sad to say, all too often the inhabitants of those so easily annexed nations had not got a clue what was happening. They allowed the invaders to step ashore. After an exchange of presents they welcomed them as friends. They allowed them to build accommodations and defenses. They allowed them to take over. Only then, I suspect, did a few of the more intelligent ones exclaim, "Ooops!"

I'm not a fan of war. I lived through too much of it as a child, even so, I have a feeling Alexander's version of expanding his nation's territory was more honest because both sides understood the consequences.

With my money belt stuffed with dollars from my mechanical endeavours the time came to leave Sudan's history behind and move on, before my visa expired. I crossed the river to Omdurman.

I roamed the dirty streets of Omdurman without seeing anything worth fighting for. Over 10,000 men, British and Africans, died in a battle to win Omdurman in the late 19th century. What a waste. So much for religious principles and foreign policy. Ten thousand dead on a wasteland of blood, guts and sand in sight of the life-giving Nile.

Hapi, the ancient male, with the body of a female, God of the Nile River and of fertility must have been outraged at the wanton desecration so close to his/her domain.

CHAPTER ELEVEN

Timbuktu was calling me. Timbuktu and the Tuareg people, and their Queen – Tin Hinan. The call was loud. The journey was long. To get there I had to cross the greater part of the southern Sahara from east to west.

My map of Africa showed the distance from Omdurman to Lake Tchad to be close to 1,200 miles. Too far to walk, I decided. I hitched a ride on a truck going to El Obeid. When we suffered a puncture I paid my fare by repairing the inner tube and the tire. The driver, a Sudanese giant – twice my size in most directions – was impressed enough by my skill that he handed me over to another trucker, Yusuf, for the next stage to El Fasher.

"He is a good man," he introduced me in his accented baritone. "He can fix your truck."

"What's wrong with his truck?" I asked.

"Is no good. Breaks down too bloody much."

He was correct. The old truck, looking as if it had been put together from half a dozen different makes, did break down too bloody much. The 370 miles proved to be a nightmare. Yusuf was not an educated man. I doubt that he had ever attended school. He knew nothing about the world; precious little about mechanics and even less about the workings of his truck. An estimated drive of two, maybe three days (the trucker's estimate, not mine), took 10 days. I spent much of that time persuading the engine to keep working for a few more hours.

Days behind the driver's schedule, as the sun was setting, I suggested we drive at night, even offering to take my turn behind the wheel. Yusuf pulled the switch for the lights. Nothing happened.

"Is no good," he shrugged his shoulders and stopped the truck. While he lit a fire and started cooking a meagre meal of rice and meat, I took the headlights apart. Both bulbs were dead. I took them out and showed Yusuf. He became quite agitated.

"That's why the lights no work," he shouted. "You take bulbs out."

"No, that's not the problem," I tried to explain. "I took the bulbs out because they don't work. They have burned out." I held them up so he could see the damaged filaments.

"You broke them?" he asked, his voice angry again.

"No, I did not break them. They were already broken. That's why the lights don't work."

Yusuf glared at me, not quite understanding. "You break bulbs? Now my lights don't work!"

"Have you any spare bulbs?" I asked.

"Why you break? Now, no bulbs. No lights," was not exactly the answer I wanted to hear.

I went back to the truck, replaced the burned-out bulbs, screwed the Perspex lenses over them and returned to Yusuf at the fire.

"Where my bulbs?" he asked.

"I put them back in the lights."

"Ah, good," he smiled at last.

The next evening, as sunset threw purple shadows over the land, Yusuf pulled on the light switch. Predictably, nothing happened. He looked at me and said, "We stop now. No lights."

Of course. What else could we do?

When we limped into El Fasher days behind schedule Yusuf punched me on the shoulder in a gesture of friendship and said, "Is a good truck, no?"

"Yeah, Good for nothing," I grimaced and punched his shoulder to return the camaraderie.

Omdurman was a dump. I remembered writing that in my journal. Compared to El Fasher, Omdurman was a cultured city. El Fasher was a dump like nothing I had ever seen. In fact, the open area where trucks were loaded and unloaded was the dump. A real dump. A garbage dump of exceptional olfactory agony. A flock of a few score miserable looking vultures watched us unload. They perched with shoulders hunched and bare necks exposed on piles of rotting rubbish. The sinister way they looked at me suggested they were considering me as the source of delicate hors d'oeuvres.

My dislike of El Fasher grew with every minute I stayed there, and the number of minutes climbed higher by the day. No trucks going west. I had heard that Amelia Earhart visited El Fasher, presumably to refuel her aeroplane, during her fatal round the world flight. What on earth, I wondered, possessed her to fly into this godforsaken piece of purgatory?

By day four I was ready to look for a ride back to Omdurman. Then a crazy, garrulous Nigerian turned El Fasher into a carnival.

He arrived at midday in a black Mercedes four-door saloon with a uniformed driver and a small pennant fluttering from the radio antenna. He sat in the back. His chauffeur, complete with a black flat-topped hat with a shiny visor, opened the door for him. Apart from the vultures, I was the only one foolish enough to be out in the heat to welcome him to El Fasher. He stepped out like visiting royalty. He was one of the largest, blackest man I have ever encountered. He wore a cream-coloured tropical suit, white shirt and what looked like a British regimental tie. His feet were encased in a pair of black shoes big enough to carry freight on the Nile. He was magnificent. He placed an oversize Panama hat on his head and greeted me as if I were an important emissary from the town.

"My name is Agha," he said, his voice rumbling from the low end of his belly: a voice so deep and rich it would have made Paul Robeson sound like a soprano. He clasped my hand and it disappeared inside the enormous paw. I wondered if I would ever get it back as he crushed my bones.

"Hi," I returned his greeting, wincing at the pain in my right hand, "I'm Tinker. Tinker Taylor."

"You are a tailor, but you don't wear a suit?" he sounded like an earthquake in full revolt.

"Yes. You are Agha. Is your last name Khan?" I was being facetious and, I suppose, rather rude. It didn't matter. My childish guess was spot on.

"Indeed it is," he smiled. "I am Agha Khan, a businessman from Lagos."

I looked at the Mercedes. I knew the car. I recognized a dent in the right front wheel hub.

"I know that car," I said. "It used to belong to the German consul in Khartoum. I overhauled the brakes on it a few weeks ago."

Agha looked at me and frowned, the lines on his massive forehead deep enough to bury complex thoughts.

"Yes," he drawled. "It is mine now. I am taking it to Lagos."

Last time I had seen the car it had German plates with a CD (Corps Diplomatique) addition. Now it had Nigerian plates. I was about to ask a question when my brain slowed my mouth. Suddenly, I didn't want to know. He continued, seeming to register what I had told him. "You are a mechanic?" he asked. "You can repair this car of mine?"

"If you have some tools, and if it needs it, I can repair it."

"You will come with me to Nigeria," he announced. I wasn't sure whether it was a question or a statement of intent. Didn't matter. Nigeria was on my route. Hallelujah!

Agha said he was in a hurry. First though, he insisted, he had business to attend to. He must also meet the mayor of El Fasher and the chief of police. I sat in front with the driver to navigate, although it was obvious the chauffeur knew the way to his destination. Agha sat in the back. After the excessive heat outside, it was cold in the airconditioned car. Followed by an increasing crowd of excited children and curious adults, we drove into the centre of town.

El Fasher's streets are laid out in a grid, separated by squares of single-storey buildings for the most part. The streets are the colour of the sand. The buildings, both public and private, are the same colour as the sand. The people vary in colour from milky coffee to dark chocolate. In between those extremes, they were the same colour as the sand. Agha, the immense black man and his colour-coordinated Mercedes, stood out like ebony beacons.

It did not occur to me at the time that I was the other end of the spectrum. My exposed skin had been darkened by months exposed to the tropical sun. Even so, I was the palest person in town.

We stopped outside a building that looked like every other building in El Fasher. Agha got out of the car carrying his briefcase. He strode the few paces to the door, which opened at his approach. It closed again as he entered. I sat with the chauffeur in airconditioned luxury for half an hour.

Agha returned without saying a word and we continued to the mayor's office. That worthy must have been the wealthiest man in El

Fasher. He emerged from his exalted space wearing a spotless pink robe that reached to his sandalled feet. A thick gold chain hung around his neck. His shaved head was bare.

He greeted Agha as a revered guest. His hands, when he extended them to Agha to shake, had rings on all fingers and both thumbs. His wristwatch appeared to be a Rolex, or a clever fake. He ignored my presence and ushered Agha into his office. The door closed behind them.

I leaned against the Mercedes with the chauffeur. He smoked a cigarette. I watched the door to the mayor's office. After a short wait the door opened wide enough for a black hand to emerge. It waved in our direction. The chauffeur understood. He opened the rear door and took out Agha's thick briefcase, which he delivered to the lone hand. The door closed again. We waited. The watching crowd grew larger.

Eventually Agha and the mayor emerged, both wreathed in smiles. Agha handed the briefcase to his chauffeur who placed it on the front passenger seat of the car – my seat, I thought. Agha and the mayor got in the back. The car moved slowly away. I followed on foot.

Behind me marched a sizeable population of little people. Every kid in El Fasher had turned out to watch the proceedings, whatever they might be. I felt like the pied piper of Hamelin as I led the chattering horde in pursuit of untold excitement. We didn't have far to go. The car stopped one sand-coloured block away in the middle of the road outside the sand-coloured police station. I put up my hand to stop the advancing army of children and adults. They obeyed with a collective sigh.

We waited. Five minutes, maybe more. The Mercedes remained closed with Agha and the mayor staying cool in its airconditioned interior. A muttering from my followers announced a change. The door of the police station opened. The Chief of Police, resplendent in a crisp sand-coloured uniform bedecked with ribbons emerged onto the step opposite the black car. He stood at attention. Last time I had seen him was when I had my passport examined and stamped on arrival. Then he had been a scruffy individual wearing dirty white shirt and grubby khaki trousers.

We, the horde of onlookers, watched and waited in the boiling sun. The Chief of Police waited in attendant heat. A car door opened. The chauffeur stepped out, closing his door behind him. He stood with one hand hovering beside the handle of the left rear door and we all waited.

Who the hell is Agha? I wondered as I fanned my face with my sand-coloured safari hat.

The chauffeur came to attention and opened the door. Behind me the crowd gasped, "Oooh!"

Agha stepped out, a smile stretching his face to show a mountain range of snow-capped teeth. The mayor slid out behind him. The chauffeur shut the door again. The engine was still running. I knew the air-conditioning would be on.

As Agha took a step forward, the Chief of Police stamped his right foot and delivered a salute that would have done credit to the guards at Buckingham Palace.

I half expected a military band to start playing. I could almost hear the staccato rhythm of the snare drums and the haunting skirl of the bagpipes. A fly settled on my nose to drink a drop of perspiration. I fanned it away with the brim of my sand-coloured safari hat. The fly circled my head, calling its family to enjoy the bounty. They came in like a squadron of spitfires. I fanned them away with... well, you know the hat by now.

Agha and the mayor reached the Chief of Police. They all shook hands and went into the station. We waited. I took a couple of paces forward. My followers followed. I turned and motioned them to stop. They did so. I walked up to the chauffeur and asked him to explain what was happening.

"I don't know," he replied.

I tried another approach. "Who is Agha?"

"He is Agha Khan," answered the chauffeur.

"I know that. But who is Agha Khan?" I'm sure the element of frustration in my voice was obvious.

"Agha Khan is a businessman from Lagos. He is very rich," intoned the chauffeur.

I gave up. By this time the horde of onlookers had seated themselves across the road. Some started singing. Others played clapping games. I heard a child crying. We waited. I was hot, much too hot, but too curious to look for shelter. My curiosity, however, was losing its edge.

The door to the police station opened. Agha stepped out followed by the mayor and then the Chief of Police. They were all smiling. My followers began to applaud with an infectious beat. Agha waved to them. They sighed, "Oooh!" as the chauffeur opened the car door and Agha eased his bulk onto the cool, air-conditioned leather of the back seat. The door closed. The chauffeur took his seat behind the steering wheel and the car slowly moved away. The crowd began to break up, most discussing the phenomenon of the big black man in the shiny black car.

The mayor and the Chief of Police stood watching as the car turned a corner and passed out of sight. I walked up to the mayor and asked about Agha.

"He is a very important man. His name is Agha Khan. He is a businessman from Lagos in Nigeria," he answered and walked back to his office. I turned to the Chief of Police, ready to ask the same question. He shrugged his shoulders before I had my mouth open and went into his office. The door slammed shut behind him.

I strolled back through the now almost empty sand-coloured streets to the sand-coloured hovel I had slept in for the past few nights. A shiny black Mercedes was parked outside. Its engine purring in anticipation. My backpack leaned against the wall of the hovel. That was a surprise. I had left it on the flat roof. As I came alongside the car a tinted window opened. Agha's voice rumbled, "We are leaving now. Do you have any clean clothes?"

I took the hint and stripped off my less-than clean shirt, wiped sweat from my armpits with it, and stuffed it in a side pocket. I knew I had a clean tee shirt near the top of my bag. I rummaged through, found it, put it on, tied up my bag, hoisted it into the now open trunk of the Mercedes, where there was a large wooden crate and expensive leather suitcases. I shut the lid and took my place in airconditioned luxury in the front passenger seat. Damn, it was cold in that car.

When we crossed the border into Tchad Agha took his briefcase into the rundown customs' building while we, the chauffeur and I, waited in the car as ordered. He came out no more than five minutes later, got back in the car and away we went.

Welcome to Tchad! No formalities for us. I could not contain my curiosity any longer. I turned in my seat and asked Agha, "Who are you?"

He laughed, a dark brown sound that made the car vibrate. "I am Agha Khan," he began, "a businessman from Lagos."

Big surprise, I thought. I've heard that answer somewhere before.

"May I ask what business you are in?" I tried polite caution.

"Oil and shipping," he rumbled. "I have oil tankers. I have cargo ships. I have many trucks. Many businesses. Here, there, everywhere." He pointed in all directions.

For the next few hours as we bounced and scraped along a poorly maintained road across semi-desert, he regaled me with tales of his exploits during his rise from a poor policeman's son in Kano to, and he said this with no hint of boasting, one of the richest men in Nigeria, maybe in Africa. Strangely, he glossed over his acquisitions of the ships.

"You seem to know the mayor and chief of police in El Fasher rather well," I said.

"It is good business to have many contacts, many business associates, in many places. I do them favours. They do me favours. It is a good system for all of us."

I got it right there. This upstanding businessman was also a crook, maybe a smuggler. He paid off the officials at every major town along his route, and at the border posts. But what was he smuggling? Whatever it was, he carried it in the wooden crate and in his briefcase, I assumed. What would he be carrying from Khartoum to Lagos?

I took a guess that he might not be quite what he claimed to be. I guessed he was, in fact, smuggling cars and laundering money – the money being in the crate. I began to suspect that he was not the legal owner of the Mercedes. My safety antennae went on full alert.

Agha talked about himself throughout the bumpy ride from El Fasher to Fort Lamy beside Lake Tchad. He talked about what we would do together when we got to Lagos. He mentioned ownership of two, maybe more, nightclubs. From his description, I assumed he meant brothels. My safety antennae went higher, and I started to plan my escape.

Across the Chari River from Fort Lamy is another country. A long finger of Cameroun points between Nigeria on the western side and Tchad on the eastern shore. The tip of the finger reaches to Lake Tchad. The only way to get where I wanted to go was across Cameroun, into Nigeria, passing through Maiduguri and Kano before taking a northerly route into Niger. First, I needed a visa for Cameroun and a visa for Nigeria. I did not tell Agha of this. He would have told me they weren't necessary.

At Fort Lamy, while Agha was paying off some officials and his chauffeur (I never did know his name. I called him Abdul and he responded) went to refuel the car, I retrieved my backpack from the car at the service station and slipped away through the convenient maze of a colourful bazaar.

Although I've never enjoyed crowds, I am happy in the hustle and bustle, the extravagant colours, the sights, sounds and smells of African and Middle Eastern bazaars. The bazaar on the edge of Fort Lamy was a classic of its kind.

Collecting the necessary visas for the next stage of my mammoth trek across Africa took six days, something of a record in the former colonial West Africa. The visa section personnel at the Cameroun Embassy were fast and efficient. They issued me a transit pass valid for two days from date of entry. That took one day. The Nigerian Embassy worked slower. I had to answer a long list of pointless questions. The attached questions are real. I watered down the answers you see below that I wanted to use.

Q. "Where are you going?"

A. "Timbuktu."

Q. "Timbuktu is not in Nigeria." (Not a question but close enough.)

A. "I know, but I have to cross part of Nigeria to get there."

Q. "How long will you stay?"

A. "Not long, I hope. Depends on getting rides. Maybe one week, maybe two."

Q. "What is your next destination (which country)?"

A. "Niger."

Q. "How will you get there?"

A. Easy answer. "Tinker go walking!"

Q. "Do you have an onward air ticket?"

A. "Me? An air ticket? You've got to be joking."

Q "Do you have sufficient funds for accommodation and meals while in Nigeria?"

A. "That's never been a problem for me."

Q. "Are you a criminal?"

A. "Depends on who you ask."

Q. "What is your occupation?"

A. "Philosopher, observer, professional traveller, bon vivant, tinker. Your choice."

Q. "Why do you want to visit Nigeria?"

A. "I don't want to visit Nigeria, but it is in the way, so I have no choice."

I kept most of my flippant replies hidden. Instead I offered a list of platitudes to express my excitement at the possibility of visiting Nigeria. I did not tell the Nigerian Embassy people that I had travelled from Sudan with the illustrious Agha Khan. A feeling in my gut told me that would not be a smart idea.

I was granted a visa for a two-week stay with departure by air. No employment allowed! The departure by air clause was a slight problem. I chose to deal with it when the time came to leave.

I was fortunate in getting a ride on a reasonably comfortable bus from Fort Lamy all the way to Kano. The paying passengers rode inside. I, and three other financially distressed African wanderers – all of them with skin many shades darker than mine, settled ourselves on the roof-rack amongst the baggage for the first uncomfortable stage across Cameroun to Maiduguri.

The noisy open-air bus station in the first Nigerian city on our schedule resembled a cattle market without the bovines. As we left for Kano, with me clinging to the roof again, with a handful of new companions, the bus faltered. It burped. It groaned. It excreted an impenetrable cloud of black smoke and then it died. I climbed down the rear ladder to the ground and told the driver, "I think your head gasket has blown. I'll repair it if you wish."

That's how to make friends. The driver raised the cover of the engine compartment and looked at his unhappy steaming motor. He looked at me.

"You can fix?" he asked. "How long?" And then, concern for his profit taking over, he asked, "How much?"

My answer was simple. "No charge. Give me a free ride inside the bus on a real seat to Kano, that's all." We shook hands and the deal was brokered.

The bus driver had a friend, or relative, who owned a garage in Maiduguri. He hitched a ride on the back of a motorcycle to borrow or buy sufficient tools and parts – I had given him a list – for me to effect repairs. He was gone all day. In his absence I slept across the back seat of the bus, the other passengers having dispersed to destinations of their own choosing. When the driver returned, I asked him why he hadn't commissioned his friend, or relative, owner of a garage, to repair the bus.

"He charges too much. He's a crook," he explained.

Another day passed while I tinkered with the engine. By the time I finished night had fallen. None of the passengers had returned. I slept on the back seat of the bus.

There's a mysterious telegraph system at work in Africa. Somehow all the passengers were rounded up and ready to board the bus at daybreak. When all were seated, the bus was full. No seat for me. I pointed out the discrepancy to the driver. He was not fazed in the least. He grabbed a young man by the shoulder, dragged him off the bus and ordered him to the roof-rack. I took his seat inside and away we went. Problem solved. The engine behaved all the way to Kano.

Not having enjoyed the luxury of a shower for some weeks – not since Khartoum, in fact – I visited the once important Kano Club. There for a small fee, I had a lukewarm shower and washed my clothes. Once they were dry, I treated myself to a cold beer beside the pool. There I heard of Agha Khan again.

"He's not Nigerian," scoffed an English member of the club. "He's Sudanese. He owns brothels in most towns and cities between Lagos and Cairo. He's a crook. He pays off the local authorities wherever he has business interests. He was probably planning to kidnap you and sell you into slavery. It happens, you know."

Now that would have been a different kind of adventure, I thought. Instead of voicing the idea, I asked, "Is his name really Agha Khan?"

"No," another scoff. "I think it's something like Mohammed Ilaouad." The Englishman, with a Yorkshire tone to his voice, pronounced it 'ill wad'. Somehow that seemed more appropriate. And so ended my association with a 'businessman from Lagos' named Agha Khan. I never heard of him again.

An overloaded truck, painted in a rainbow of colours, carried me to Katsina. Another overloaded truck of similar artistic expression took me to the border at Jibiya and there I ran into a stumbling block.

"Your visa is only valid for air travel out of Nigeria," the border commander told me.

"Okay," I smiled at him. "Where's the nearest airport?"

"Kano," he replied, pointing south.

"Are there flights from Kano to Niamey, or Maradi?" I asked, knowing the answer was negative.

"No. I don't know," he admitted. Gotcha! I thought.

"Well, Niger is only five minutes away on that road," I said. "Wouldn't it be easier if you sent me across the border on foot? I can't walk all the way back to Kano."

He scratched an armpit; looked me up and down. "You have cigarettes?" he asked. An interesting non sequitur.

"No, sorry. I don't smoke."

He studied my passport from cover to cover. Returning to my visa, he said, "Your visa says you must leave by air."

"I don't know why that's written there. I can't afford to fly."

With a sigh that could have been heard in Lagos he sorted through an untidy collection of wooden handled rubber stamps until he found one he liked. Breathing on it twice, he stamped it hard on my visa. Handing the passport back to me, he said, "You can go. Next time you fly!"

I walked the short distance to the Niger border post where the truck I had been on was held up. An hour later I was on top again and on my way to Maradi. A French army officer driving a pickup truck gave me a ride all the way from there to Niamey. He said he was based at Zinder, well to the east of Maradi. Only later did it occur to me that he must have been with the French Foreign Legion. I didn't like him. He treated the locals we met with the arrogance and disdain exhibited by many colonials. I appreciated the ride. Even so, I was happy to leave him in Niamey.

Seeing some faded postcards on a spindly rack outside a shop, I bought one. I hadn't written home since Khartoum. Now seemed like a good idea to tell my family I was alive and well.

The fastest route from Niamey to Timbuktu, should rides be available, would be north to Gao, in Mali, and then follow the Niger River past Bourem and then west to Timbuktu. That would mean relying on a rarely used sand track of at least 200 miles between Bourem and Timbuktu. Not such a good idea.

Local advice as to the best route came from conflicting sources. Some said go to Gao and get a riverboat from there, or a truck going west to Bamako. Others recommended a route through Upper Volta, longer but more scenic. The decision was made for me when I heard of a truck going to Ouagadougou, capital of Upper Volta. Although I was anxious to see Timbuktu, I also wanted to visit the Dogon people who lived on the Bandiagara escarpment in Mali, south of Mopti. Ouagadougou was on the way.

The drive took us through the bushlands just north of the W game reserve, which spreads over parts of three nations. To my delight we passed a troop of monkeys chattering among themselves in a tree. Soon after we watched as an elephant blocked the road so it could browse on moist green leaves. I stood up on top of the truck to see it better. Our driver ruined the show by pounding on the truck's horn. The blaring noise annoyed the elephant as much as it annoyed me. Flicking its great ears and swinging its trunk, the pachyderm pushed into the bushes and cleared the road.

I met a woman in Ouagadougou, or in truth, she met me. I was buying bread from a market stall when she thundered through like 300 pounds of battleship looking for trouble. Dressed in a voluminous floor-length dress of many colours, and a turban to match, she was about ten years my senior and almost as dark as Agha Khan. She accosted me, albeit with a charming smile, and demanded to know my name and where I came from. My answers were sufficient to earn an invitation to her bar. I said, "No, thanks. I have no money for beer."

"No pay," she insisted, with a smile and a fluttering of eyelashes. "I am Chantal. You are my honoured guest."

Taking me by the arm with a grip like a vice, she led me through the market past stalls selling meat, others selling fruit, all of them buzzing with flies, and along a dusty side street. Not once did she stop talking and rarely did she take a breath. In her bar, which was devoid of drinkers, except for an old man in a corner by himself, she sat me down at a low table.

"My papa," she explained, pointing to the old man as she poured me a beer. I took a swallow, finding the beer well chilled. I began to think what a lucky fellow I was to have found a new friend. Chantal sat opposite me and pulled her skirts up to just above her ample knees. She lifted one bare foot onto the edge of the table. It was not an elegant way of sitting for a lady. Her legs were (I'm being polite here) rather large. Madame noticed my glance in the direction of her lower limbs. She smiled, almost coquettish, and slowly separated her knees.

"You like?" she asked, leaning forward after a few seconds to display a bosom the size of a pair of prize watermelons.

I stared in a mix of wonder and horror, my eyes flickering from her enormous bust to the darkness of infinity and back again. She opened her legs wider. That was too much for me. I had no wish to share her carnal dreams. She was too blatant; too pushy. She gave me the impression she liked to be on top. With my slender frame I would have been squashed before... well, you understand. I scattered a few coins on the counter, grabbed my pack and I ran, leaving the remains of my beer to lose its chill and its fizz on the table.

Fear of being found again by Chantal forced me to flee Ouagadougou. At last I understood and agreed with Falstaff's words in Shakespeare's *Henry IV*, "The better part of valour is discretion." I caught the first truck north to Ouahigouya.

My only impression of Ouahigouya is that it reminded me of El Fasher, both for the quantity of its vultures and its complete lack of attractive features. The long dirt road from there to the border with Mali looked as dry and dusty as Ouahigouya's streets and infinitely more interesting. I started walking at daybreak.

Two cars, empty apart from a driver in each, passed me in a cloud of dust. I cursed them both, for the dust and for not stopping. By midday, with the sun high and hot, I had to seek shelter under a tree, but within easy reach of the road. I waited a long time. As the afternoon began to wane a taxi-en-brousse came in sight. I made the driver stop by the simple expedient of standing in its path and waving my arms. It could only be going to the border. I intended going with it.

To say it was crowded would be an understatement. With much laughter and help from the existing passengers, I squeezed in holding my pack on my lap. A rotund lady beside me smiled and offered me a mango. She reminded me of Chantal. I almost panicked.

The taxi, in reality a small pick-up truck with two bench seats in the back and a thin canvas roof, would have been crowded with three people on each side. When I boarded I counted heads and noted that I was number thirteen in the back, three more were in the front with the driver. Crammed together as we were, intimacy was inevitable. I had half a hefty hip draped over my thigh on one side and a squalling child with a very dirty nose almost on my lap on the other. Although the back of the truck was open, the smell inside was trapped by an excess of sweaty humanity. Chantal's double offered me the mango again. I accepted graciously. I took one bite, savoured the sweet juice, and then made the mistake of looking to my right. The snot-nosed baby reached out dirty hands for the mango. My appetite disappeared. I gave the kid the mango. He massaged the fruit with fingers, teeth and gums and dribbled on my leg for the next eternity.

At the border everyone was passed through without question. Everyone that is except me. I was invited to get out of the taxi with my pack and ushered into a hut hot enough to have doubled as a sauna. There I was told to wait. I leaned against a wall and felt the sweat trickling down my spine while two border guards took it in turns to read all the entries on all the pages in my passport. They then discussed most of them. I heard an engine start. A vehicle left a cloud of dust hovering over the border post as it swayed along the dirt road to its next stop somewhere in Mali. I knew it was my ride.

"Quel dommage," commented one of the men. "Votre taxi, il a parti." What a pity. Your taxi has gone.

My mind wanted to reply with Anglo Saxon vehemence. Fuck you, it thought. My mouth, having learned much during my travels, stayed silent.

I spent that night and half the next day at the insignificant border post with two insignificant officials. I was not in a good mood. The first taxi-en-brousse that came along had no choice but to fit me in with its already compressed passengers. I yelled to the driver that I was going to Bandiagara. He understood and shouted the fare, equivalent to one dollar.

"Oui, d'accord," I shouted my agreement.

At Dourou, on the Bandiagara plateau, close to the escarpment, I found a small room in which to spread my sleeping bag and to leave my backpack in safety. A boy about ten-years-old lived there with his parents and two smaller children. The boy, who said his name was Bemba, offered to be my guide to the dwellings on and in the cliffs. Bemba was a bright kid, although he spoke only a smattering of French and knew nothing of English. We communicated in a confusing mélange of Dogon, French, Arabic and sign language. We understood each other some of the time.

On our day-long excursion, Bemba was most proud to show me a site used for circumcision rights. None were happening that day (for which I was thankful). Bemba showed remarkable acting talent when he pantomimed the male circumcision operation — alternating being the witchdoctor with a sharpened stick and the victim. His graphic few moments of painful theatre brought tears to my eyes — I could almost feel the sharp pain. He completed his act by displaying his own immature circumcised penis to show me the end result (forgive the unfortunate pun, please), a finale I could have done without. Then he asked me about my own experience. At least, I think that was the question.

I growled, "None of your business."

Bemba said something incomprehensible, shrugged his narrow shoulders and led me on to the next experience. That entailed scrambling down a narrow defile through the rocks.

"Be careful here," he warned, pointing to loose rocks.

Bemba's caution was justified. The narrow path down the escarpment to a collection of houses and other buildings was dangerous, but the descent was worth the risk. We came upon tiny houses set into the cliffs. Constructed from the same materials as the backdrop against which they stood, the collections of little thatched-roof dwellings were not always easy to see. From the plain below the Bandiagara escarpment they would have been all but invisible. I have seen some unusual dwellings in my travels. The homes and the granaries of the Dogon people were like something out of a faery story. And they weren't much bigger.

The Dogons are mostly animists, with a few Christians among them. They rejected the arrival of Islam and built their villages in the most inaccessible places to deter enemies — including Islamic missionaries. Few that I met spoke any French. They conversed in their own Dogon language.

Bemba led me back up the cliff and to his home where, tired out from my exertions, I soon fell asleep. In the morning he directed me to a dusty track where I might be able to find transport. I did. I hitched a ride with a small group of French tourists heading for Mopti.

The river front at Mopti is a sight not to be missed by West African travellers. Many of the Fulani women wear huge gold earrings, so heavy that they had to be supported by a braided band of leather, or hair, over their heads. Many were so heavy that they had stretched the ear lobes to unsightly lengths. Gold-decorated Fulani women, some beautiful, some not so appealing, bargaining, buying, selling. Men, also Fulani, I assumed, unloaded a large wooden boat carrying flat bars of salt from the northern mines. I would see more of that at Timbuktu.

An unhappy camel spat a stream of saliva at three small boys as it was led over the slippery flat rocks where boats were unloaded. Its owner lifted his dirty robe and tucked it up between his legs. Bare from mid-thigh down, he dragged his unwilling camel into the river. Was he trying to drown it? I wondered as I watched in fascination. No, the object of the river visit was to wash the camel. At least that was the man's intention. The camel disagreed, screaming epithets to the waterfront about the harshness of a life near a river.

I suspected the camel might have seen crocodiles in action somewhere else and was just expressing its desire for self-preservation. The man was not deterred by his camel's complaints. He scooped a handful of water and rubbed it over the camel's muzzle. Teeth flashed. The man swore. The camel snarled. A helpful boy offered the man the use of a large blue plastic bucket. I moved so as not to receive an unplanned shower.

The man had a problem. He needed to use two hands to fill the bucket and throw it over the camel. To do so, he would have to let go of the reins. That would free the camel which was, I believed, unlikely to stay long enough to get wet. The man solved his problem by standing with one foot on the end of the reins. The camel watched, flicking its long eyelashes at a few flies. The man took his eyes off the camel to fill the bucket. The camel, being possessed of natural intelligence, seized the opportunity. It swung its great head to one side, tearing the reins away from under the man's foot. Free, the camel turned, pushed its owner off balance into the river and raced for dry ground.

Laughing out loud, like most of the other onlookers, I watched as the now soaked camel owner raced after his beast waving the borrowed blue bucket in the air and shouting dreadful imprecations. A phalanx of young boys, one the owner of the blue bucket, raced after the wet man chasing the dry camel through a crowded river port. I watched, still laughing with everyone else, until the impromptu comedy act had disappeared towards the mosque.

Getting to Timbuktu from Mopti by road meant a huge detour, if a ride, or rides, could be found. The only sensible course was to travel by river. That would cost money. I walked the waterfront, going from boat to boat, asking who was going downstream to Kabara, the river port for Timbuktu.

The boatmen saw a white man, assumed I was French and the price of a ticket soared. I explained that I was a poor man from Canada, assuming they would consider an Englishman to be as rich as a

Frenchman. That didn't work. The first boatman I tried it on exclaimed, "Ah, Canada America. Very good. Very much money." The price rose higher. I went in search of cheaper transport.

There was a service from Koulikoro by river steamer that called at Djenne and Mopti en route to Kabara and Gao. I visited the appropriate ticket office in Mopti and was told that French people had to travel either second or first class. Third class was only for natives. I explained that I was not French. He amended the wording of the rule from French to European. The second-class fare was more than I could afford. I returned to searching the waterfront for a less expensive option.

A group of women, young and old, stood in the shallows washing their clothes, laughing and talking in loud voices. Many were topless: some young and firm, others varying from saggy to unmentionable. Children played in the water close by. Seeing more boats off to the west, I walked the riverbank to where an old man sat with a long-shaft outboard motor between his legs. He was hitting the top of it with a heavy wrench. I crouched beside him. He muttered threats and curses as he assaulted the engine.

I put my hand on his arm and took the wrench from him, shaking my head in disapproval. He frowned. I signed that I could fix the engine. He pushed the outboard motor towards me as if to say, "Go ahead."

I stood up. "Attends," I ordered. Wait.

Half an hour of begging and borrowing later I had a set of small tools in my hand. Finding a cleanish sheet of plastic, I laid the motor on it and began to look for problems. The spark plug needed replacing. That was easy. The cylinder head was not as tight as it should be. No doubt the bashing with the heavy wrench had not helped it keep its integrity. I tightened all the bolts. I cleaned everything. I filed rough edges from the propeller blades, and I sent a young boy to buy a new sparkplug.

When the motor was ready I lifted it into the boat and climbed aboard. Watched by the old man, I fitted the motor to its bracket near the stern of his pirogue and clamped it tight. Three pulls on the starter cord had the engine spluttering. The old man stood up, a smile playing around his mouth. I made a couple of adjustments and pulled the cord again. The outboard came to life. The old man was delighted. He scrambled aboard as I stopped the engine. Then, with a confidant grin at me, he pulled the cord and the engine started again. He laughed and patted the cylinder head.

"Bon," he said. Good. He shook my hand, his lined face wrinkling with his smile of pure joy.

The old man was happy. He could go fishing again. He and his family would eat that night. Nothing else had changed on the Mopti riverside. I waved goodbye and continued my search. The following day, while I was sipping tea at an outside stall, a young man approached.

"You speak Engleesh?" he asked.

"Yes, and I speak some French, if you prefer."

"You help the old man yesterday," he said, pointing towards the river. "He is my grandfather. Where do you want to go?"

"Tombouctou," I answered using the French version.

"Come," he said. "I arrange for you."

We walked together to one of the large wooden boats. The owner, or captain, stood close talking with two men. He had quoted me an outrageous price the previous day after the salt had been unloaded from his craft. My benefactor interrupted the conversation. Speaking in a rapid dialect which I assumed was Bamanakan, the Bambara language spoken by most people in Mali, he negotiated for a place in the boat for me. The discussion, which included much pointing at me, and to the old man out fishing, lasted long enough for me to repair two engines.

Finally, the young man addressed me. "He says he is leaving tomorrow morning. Can you pay him now to reserve your place?"

"How much?" I asked.

The fare was equivalent to about five dollars. I agreed but refused to pay until I boarded the boat. That caused another few minutes of argument.

"Okay," agreed the boatman at last. "You can pay me now and sleep on boat tonight, but you must bring your own food."

That worked for me. I thanked them both, paid the boatman, hefted my pack onto the boat and went shopping for bread, dates and mangoes: the cheapest foods available.

The Niger is the third longest river in Africa, exceeded only by the Zambesi and the Nile. My boat voyage from Mopti to Kabara would cover more than 180 miles and take me through the internal delta where the Niger spreads its waters into lakes and connecting streams.

We departed Mopti as the stars began to fade. I sat near the bow, as far away from the engine as possible. From my prominent position, I hoped to see many species of the Niger's wildlife.

The boat was crowded with local people, each of whom would have paid somewhat less than I, seated on top of a cargo consisting of rice, millet, peanuts and cotton. I had tied my pack to cargo in the bow, which the boatman had told me would be unloaded at Kabara.

The river was busy with fishermen and their nets. Our captain steered his much larger craft with great skill between all potential obstacles. One fisherman held up a large Nile perch, known as *Capitaine*. Our captain bargained for it as we passed. Money changed hands and the fish, about as long as my arm, thumped onto the deck at his feet. The captain would have fish for dinner that night. I had bread and dates to look forward to.

Travelling by boat is relaxing. Much as I enjoyed walking, the river soothed me into wishing I could cruise on a boat forever. During the heat of the day the Niger was quiet. All on our boat, apart from the captain and myself, dozed. He steered his boat. I watched two sides of the river and the way ahead.

As the afternoon looked forward to the night, birds began to appear skimming low over the surface of the river. Among them were plovers, kingfishers and sandpipers. In the shallows, herons risked life and limb to catch small fish while crocodiles watched and waited for a quick snack.

A mouth-watering aroma of fried fish drifted to me in the early evening. I looked back to where the captain was eating the *capitaine* for dinner while another man steered the boat. The captain glanced in my direction and motioned me to join him.

"Mangez," he invited me to share his fish. Eat. I sat opposite him and tucked in. Dinner that night was excellent.

Almost through the inland delta, after three days on the river, we stopped at Niafounké on the north bank. A gang of porters came on board, hustling the passengers ashore so the work of unloading could begin. Aided by the captain and a helper I had not seen until then, the work took a hot two hours. Some people left us at Niafounké – they were going to Goundam, I was told. The departing passengers were soon replaced by another collection of travellers going to Timbuktu and points further east.

We left in the dark and spent one more night on the river. Before daybreak we pulled into Kabara. The waterfront was crowded with slim wooden boats and with people. I was the first passenger off the boat. I grinned at everyone in delight. Timbuktu was no more than five miles to the north. Not much of a walk for Tinker.

Timbuktu

TINKER GO WALKING!

CHAPTER TWELVE

Timbuktu! The culmination of a dream to me. I walked into town, dusty, dirty, hungry, thirsty and with my world on my back. I looked around me in awe. I was in Timbuktu! The fabled Timbuktu!

Among the khaki houses in the sandy streets a crowd of half-dressed kids soon surrounded me all chattering and asking questions.

"Francais?" they asked in their thin, piping voices. "Vous êtes Francais?"

I pulled at the corners of my eyes and answered, "Non, je suis Chinois."

The kids laughed with delight. They mimicked my action with their own eyes, then returned to the refrain.

"Vous êtes Francais?"

"Non, je suis Anglais."

"C'est vrai? Ah, bon," one of them called out. "English good."

"Tombouctou magnifique," I returned the compliment and received the expected broad smiles in return.

I found lodging on the flat rooftop of a private house for a few cents a night. I had access to water and a primitive toilet. Food was never a problem. Three meals a day, not much variety, but filling anyway, cost me a few cents more. I lived on rice, dates, beans, and stringy meat – goat or camel, sometimes tripe (which I detest), occasionally fish from the Niger River (which I enjoyed very much), all spiced to perfection and washed down with hot tea.

By this time, after more than a year roaming the Middle East and the northern half of Africa, I considered myself immune to any stomach disorders. In fact, I had not had any physical upsets since leaving Austria the previous year. I could and did eat almost anything, but I often wondered what had been in that sausage I ate in Vienna.

Alone, but not lonely, I roamed the sandy streets of Timbuktu. There was so much history to discover. History that was not taught in Canadian, American, or British schools. Much of it, I suspect, not taught in French schools either.

Kankan Musa, said to be the richest man in the world at the peak of his existence in the 14th century, was the leader of the massive Ghanaian Empire of which modern day Mali was but a part. He was responsible for developing the first university building in Timbuktu – the Mosque of Djinguiraiber – and other madrassahs.

The Mosque of Djinguiraiber

Djinguiraiber was built between 1325 and 1327 of local earth, limestone and a mud substance, all held together with wooden spikes that continue to decorate the exterior of the minaret. The Mosque of Sonkoré was constructed in the same fashion, while the third of the trio – the Mosque of Sidi Yahiya, completed in 1440, is of a more substantial brick construction. Even so, all three have lasted the centuries thanks to the care of local religious leaders. Those three religious buildings formed the Islamic University of Timbuktu. Scholars and students came from all across the Arab world to study in Timbuktu.

Alexander Gordon Lang, a Major in the British army, was the first known European to see Timbuktu. On a mission for the British government, he travelled across the Sahara from the Libyan coast to reach Timbuktu in 1826. A strange man, he left Emma, his young wife of only a few days, to embark on his quest to reach Timbuktu. Emma Warrington was the daughter of the British Consul in Tripoli. She didn't know that the handsome, gallant major was a magnet for trouble. Either he found it or trouble found him. He and his nomadic companions were attacked by marauding Tuareg near In Salah. During that vicious fight Laing suffered in excess of twenty sword wounds, one of which maimed one of his hands.

The attack did not deter Laing from his quest. Showing his mettle, his determination and a wide stubborn streak, he and his surviving men continued south to the Niger River. Following local advice, they headed east along the north shore until a few high buildings came in sight.

Laing arrived in Timbuktu in August 1826 and spent a month in the legendary city. He must have known that the Tuareg attack near In Salah had been because he was a European. But nothing changed. Laing roamed the streets of Timbuktu wearing his British army uniform and openly drawing a map of the area. Not surprising then that the locals decided he was a spy. I'm sure he had never heard the expression "a slow learner" but the appellation seems to suit him. Now, perhaps, we should add another adjective to explain the defect in Laing's personality: 'Stupid' leaps to mind, but that might be unfair. 'Blind arrogance' will do, as events proved.

When he departed Timbuktu in late September 1826, in his colourful British army uniform, of course, Laing had planned to travel north via the village of Araouane and the salt mines of Taoudenit and return to Emma in Tripoli. He didn't make it. He either had a death wish or he was indeed blindly arrogant. Laing travelled in the company of Tuareg guides who, predictably, murdered him a couple of days north of Timbuktu. Emma Laing, née Warrington, was a widow but she wouldn't learn that sad fact for many months. Laing's maps and journals, which he would certainly have carried with him, have never been found.

Considering those responsible for Laing's murder came from Timbuktu, there is a certain sad irony in the fact that the house he rented for his month-long sojourn now bears a plaque over the lintel inscribed with his name and significant dates. History has turned the murdered army officer/explorer into a tourist attraction.

Likewise, the Frenchman, René Caillé, who arrived in Timbuktu two years after Laing. Caillé disguised himself as a Muslim travelling to Egypt. His disguise almost certainly saved his life as he was never suspected of being a Christian. The fact that he spoke Arabic and understood the Islamic religion also helped keep him safe. He stayed long enough to draw an incorrect but otherwise attractive sketch of the city's layout, and to dispel the myth that Timbuktu was a city rich in gold. He wrote of the dusty, sand-coloured streets and houses:

> I found it neither so large nor so populous as I had expected. Its commerce is not so considerable as fame has reported. There was not as at Jenné [Djenné] a concourse of strangers from all parts of the Soudan. I saw in the streets of Timbuctoo only the camels, which had arrived from Cabra [Kabara] laden with the merchandise of the flotilla, a few groups of the inhabitants sitting on mats, conversing together, and Moors lying asleep in the shade before their doors. In a word everything had a dull appearance."

Caillé only stayed two weeks before he joined a camel caravan departing for Fez, hundreds of miles away across the Sahara in Morocco. En route the caravan stopped at Araouane, 180 miles along the trail, no doubt to water the camels at the deep sweetwater well. Caillé was not impressed with the small remote sand-coloured village. He described it as, "The most despicable place on earth."

Poor old René was thought a liar when he eventually returned to France. No one in authority wanted to believe his description of the sordid, dirty, dusty desert town. Legend said Timbuktu was a wealthy city where gold was commonplace. It was a city of great learning. That was the legend. That was what people wanted to hear. Caillé didn't see any gold during his brief hiatus in Timbuktu and he said so. He apparently noticed nothing that resembled buildings for education, and he said so.

He wasn't believed, however he did not go empty handed. He was awarded a prize of 9,000 Francs from the French Geographical Society for being the first European to reach Timbuktu, even though that honour should have gone, posthumously, to the foolhardy Major Alexander Gordon Laing. Maybe René won the prize because he survived to tell his tale, even if he wasn't believed in all quarters.

René Caillé also has a plaque above a lintel on a house in Timbuktu. A tourist attraction, yes, but not so ironic as Laing's memorial.

A third European wandered into town in 1854, this time from Germany. Heinrich Barth was a scholar as well as an explorer. Multilingual, scientific in his approach to life, Barth had been roaming the fringes of the Sahara for five years. As befitting his station as a recognized scholar, Barth's description of Timbuktu as nothing more than a small slave-trading town on the edge of the desert was believed.

It is not surprising that the Malien government tourism folks have seen fit to award Heinrich Barth his own plaque above a lintel on a house in Timbuktu to join the other two. A trio of quite different men from different countries and different backgrounds, all honoured in the same way in the mysterious city of Timbuktu.

Tuareg, the fabled Blue Men of the Sahara – also called the 'Lords of the Sahara', are regularly seen in Timbuktu. Believed to have come from Berber stock, perhaps from the Atlas Mountains, they are pastoral nomads with the vast expanse of the Sahara as their domain. They also have a fearsome reputation as warriors. A reminder of that status is exemplified in the proud way they carry themselves, and in the way they dress. Apart from their blue robes and indigo veils, all men carry a long sword, called a takouba, at their waist. Many have at least one wrist dagger tucked up their sleeve, another concealed in their waistband, and some carry a spear for good measure. Tuareg is the plural. Targui is the singular for a man and Targuia the singular for a woman.

One afternoon, when many were avoiding the heat of the day, a teenage boy approached me where I sat writing in the shade of a palm tree and said, "M'sieu. Les chameaux. Beaucoup des chameaux." He pointed towards the north of town. "Viens avec moi," he added, pulling at my arm.

Catching the excitement, I put away my notebook and pen and went with him. I guessed what was happening and hoped I was correct. We trotted past the last of the rounded palm frond covered houses and up a dune to the top where a grove of acacia trees formed a partial wind break.

"Regardez là," the boy said, pointing to an indistinct shape shimmering in the haze.

As it came closer the shape took on the form of a man leading a loaded camel. Behind them, in a long line stretching far back among the acacias, followed another thirty or so camels and a handful of men.

The procession came to a halt in sight of Timbuktu's rooftops. The men couched the camels and began unloading two flat bars of salt from each side of each beast. This then was the *azalai*, a salt caravan from the open-pit mines of Taoudenit. Watching the nomads lifting the heavy, flat rectangles of salt from the herd of complaining camels made me think of changing plans. Should I go to the distant salt mines at Taoudenit, some 420 miles to the north, by camel, or should I continue in search of Tin Hinan? I chose the queen, of course. It was the right decision. Little did I know that two decades in the future I would organize and lead a filming expedition for CBC-TV along those desolate caravan trails to those very same salt mines.

The salt mines at Taoudenit have been worked by manual labour for centuries. The salt plain stretches across northern Mali from Terhaza in the north-east to beyond Taoudenit, over 100 miles to the south-west. The miners work in pairs by day and live in rough buildings made from the salty rubble cast out of the mines by night. The work is hard and few miners stay longer than three months before returning to Timbuktu for a rest and a more civilized life. Bars of salt carved from the ground are trimmed to exact measurements before being carried south by the camel caravans. To retain their market value, the bars must travel without suffering damage.

The camels rested that night, as did their drovers. I slept out near them, determined not to miss a moment of the action. In the morning the men re-loaded the camels and, passing to the east of Timbuktu, led the caravan to Kabara on the north bank of the Niger River. As soon as the camels felt the salt lifted from them for the last time, they rocked to an upright position and ambled away from the boats to drink in thirsty gulps from the river before being led back to Timbuktu.

Azalai – a salt caravan from Taoudenit

Now a team of men loaded the salt, bar by precious bar, onto motor-powered wooden boats. As each boat was loaded it turned into the current and set off upstream on the Niger River to Mopti – where I had been not long before. There it would be sold at the riverside market. Only then would it be broken into smaller pieces.

The locals believed the salt had magical properties, being an inexpensive and guaranteed cure for anything from syphilis to the common cold. I had my doubts, knowing it was, in reality, nothing more than sodium chloride (*NaCl*), precious enough in itself as a food additive.

While I was playing soccer with a group of small boys one afternoon I noticed we were being watched by a Targui. When I looked at him and indicated he should join us, he adjusted his dark veil and turned away. I met him again later. He came looking for me and asked if we could talk.

He introduced himself as Ali Muhammed, a Targui from the village of Ber, a few miles east of Timbuktu. He was about my age and spoke good French plus a little English. We drank tea together and we talked about our lives for many hours. That was the beginning of a friendship that would last more than two decades.

Ali Muhammed

Ali Muhammed took me to a local 'tailor' to have appropriate desert clothes made. Thanks to his bargaining skills, I was soon the proud owner of a complete set of Tuareg regalia. He also helped me bargain for a takouba to accompany the clothes. Ali Muhammed introduced me to his camel and taught me to mount safely and to ride in comfort.

Excellent lessons that I would need soon. Within a few years Ali Muhammed had become the official Timbuktu region guide for the Malien Tourism Office.

Wherever I went in the city I asked about Tin Hinan, Queen of the Tuareg. I dropped her name in the market, beside bread-baking ovens where women congregated in the early morning, and among the nomads. Many had no idea who she was. The older and better educated among the people were the ones who knew of the Tuareg queen. They greeted my questions with smiles.

"Where is she buried?" I asked of a group of old men sitting around an open fire sipping strong, sweet tea. I knew the answer was Abalessa in southern Algeria. I wondered if the Tuareg of Timbuktu also knew.

One man answered for all. "Baïd," he told me, pointing to the east. Far away. As a general direction, he was right.

"How far?" I probed.

"Perhaps twenty days," he said. "Maybe more." The others nodded and murmured their assent. Twenty days, at least. He meant twenty days by camel: the only way to travel. They knew where she was buried alright.

After I had explored Timbuktu from all angles, studied its history in the mosques and the ancient museum, visited the said houses of the earliest European explorers, enjoyed its food, and delighted in its people, I moved on. I looked back at the sand-coloured buildings as I walked south to the Niger River. I knew, deep inside, I would return to Timbuktu again and again. And, as my life progressed, I did.

TINKER GO WALKING!

CHAPTER THIRTEEN

Another long, slender riverboat took me from Kabara to the even older riverside desert town of Gao. I sat on the sharp, pointed bow under a low awning with my feet barely above the muddy water and watched the Niger River and the distant banks for signs of wildlife. The boatman and his son took turns steering the craft, which they called a pinnace. Often the only sound to disturb the soporific susurration of water against the bow was the distant put-put of the outboard motor way back at the stern

"Crocodile," the boatman called in mid-afternoon and pointed with one hand. My feet came up to deck lever without being told to. The croc, a big one, nosed towards us with malicious intent. The boatman whacked it on its gnarly snout with a stick as we passed. The croc sunk into the depths in search of less boisterous prey. My feet lowered to water level again. The river settled down. All became quiet. The only natural sounds being the occasional fluttering of wings as birds kept the river's surface clear of insects.

That night, as we moved downstream with the current, a bellowing from the south shore intruded on my thoughts.

"Qu'est que c'est cà?" I asked. What is that?

"Hippo, la bas," the son pointed to our right. The beast was too far away to offer us harm. We put-putted on through the night and through the humidity of most of the next day. I slept. I watched the river.

I thought about the places I had been and the places ahead. I thought about Queen Tin Hinan, and Cleopatra and I wondered if I would one day meet a live, modern version of either one.

We stopped three times: at Gourma Rharous, at Bamba and at Bourem. At each stop we unloaded some cargo and took on more. At each stop I went ashore to stretch my legs and look around the village. At Bourem we added a family: two adults and four kids. From the moment they arrived on board, the junior members of the little tribe watched my every move with fascination.

"Où est il?" one of them asked of their father.

"Je ne sais pas," he replied, shaking his head. He asked the boatman about me, learning no more than that I was, "Un Anglais, peut être."

A herd of giraffe grazed on the trees by the river south of Bourem. They were the first I had seen. The boatman pointed to the giraffes and then to the other side of the river, far away.

"Giraffe ici. Hippo la bas. Les lions aussi," he shouted to me. Giraffe here. Hippos over there. Lions too.

The waterfront at Gao came in sight late on the fourth day after three nights on the river. Showing skill and a certain panache, the boatman thrust the bow of his pinnace between a host of much smaller pirogues with a screech of wood on wood and came to halt.

"Et voilà," he announced with riverine pride. "Nous avons arrivè à Gao."

I shouldered my pack, stepped off the bow onto slippery wet stones, waved my thanks to the boatman and his son and walked into Gao. Evidence of the city's earliest days are scattered along the waterfront in the form of fishing boats of many shapes and sizes. Fishermen established the first settlement at this location in the 7th century, though they called their village KawKaw. I cannot imagine why they named it so and no one in Gao could tell me.

The pyramid shaped tomb of Askia Mohammed, king of the Songhai Empire dates back to 1495. As with the mosques of Djinguiraiber and Sankore in Timbuktu, it is a mud-brick structure decorated with a multitude of sun-blackened sticks protruding from its walls. The hardwood sticks, from the atil tree (*Maerua crassifolia*), are there as a permanent form of scaffolding. The pyramid is said to be over 50 feet in height and is attended by two mosques. I found the complex to be the only buildings of architectural interest in the city. But then, I wasn't there to study architecture. I was on my way to meet a desert queen.

Gao was the terminal point of a major Trans-Saharan caravan route; later – including during my visit – the end of the truck route across the Tanezrouft. Translated variously as 'land of terror' or 'barren as the palm of your hand' the Tanezrouft has a fearsome reputation as the most dangerous region in the Sahara.

There is an old camel caravan trail between Gao and Abalessa, but no longer any caravans. No trucks, either. The camel trail I hoped to follow started from Gao and followed the main route north as far as Anefis. There, I had been told, it branched off to the north-east on an unmarked track. Without the all-important caravans though, my options were limited: take a long circuitous route on a series of trucks through Niger and then north to the Ahaggar Mountains of southern Algeria, or, hmm, go the direct route by camel. The truck route would take many days, perhaps weeks, with no guarantee of transport north from Tahoua or Agadir. The camel trail looked to be a better prospect – if I could beg, borrow, rent or steal a few camels and a guide.

Renting camels is not the easiest job I have ever attempted. The Imam of the Askia Mosque sent me to his brother saying, "He has many fine camels. Give best price."

Well, that wasn't quite true. His 'brother', Mahmoud, obviously from a Tuareg tribe, unlike the Imam, did have some fine camels. He also had some that had trekked through too many sandstorms and seen a famine or two. I was wearing my recently acquired Tuareg robes, hoping to fit into the scenery. We sat down in his small sand-coloured courtyard drinking the inevitable tea and began the discussion in a mixture of French, English and Tamachek.

"How many camels you want?" he asked.

"I need five. Three for riding, two for baggage. And I need two men as guides: two Tuareg, you and one other."

Mahmoud sucked at his teeth and made notes in the sand. He grunted softly and he groaned. I waited in silence, praying for a reasonable offer. I was aware we were being watched. Glancing to my left I saw a tall, slim, attractive young lady with a headscarf but no veil. Standing in a doorway, she smiled at me. I smiled back. Her face beamed. So did mine, I'm sure. I returned my attention to Mahmoud just as he looked up at me.

"Aywah," he exclaimed in a soft rush of tea anointed breath. "Aywah," he said again. He poured more tea for both of us. Sipped noisily and licked his lips. "Aywah," he repeated.

And then he began a sob story about poverty and the cost of living in Gao. I listened in silence, nodding occasionally at some less than profound comment. Mahmoud described his life, his three wives, his countless children, especially a daughter of marriageable age, but not yet betrothed; the hardships of living in a desert community. He cited the cost of feed for the camels. He complained about a bad back, about fever, about the dangers of the trail – "Much hot. Many snakes and scorpions."

Mahmoud hesitated. He took a breath. Then, with perfect timing, he named his price, "For the camels, ten dollars each for every day. For myself and my cousin, ten dollars each for every day."

I added up the cost in my mind working on the minimum of a three-week journey. It came to nearly fifteen hundred dollars. I shook my head.

"Impossible," I said. "Impossible!"

Mahmoud gave me a shrewd look. "How much you pay?"

"One hundred and fifty dollars. That's all. Finish," I offered.

Mahmoud looked as if he might burst into tears. He moaned and he wailed. He swayed from side to side and he rocked back and forth.

"I am a poor man," he sobbed.

I reached out and took his hand, holding it in my right. Adopting my most earnest expression, I commiserated with him.

"I know you are a poor man, Mahmoud, my friend. Your three wives must cost a lot to feed and so must your children. I understand how hard your life is, but I am a poor man too. I do not have enough money to pay you what you ask, only what I offer."

He dropped his 'best' price to one thousand dollars for everything. I shook my head. "Too much."

In return I again offered one hundred and fifty dollars. He looked as though he would faint at my audacity.

He countered with five hundred dollars. I shook my head again. "No. Still too much. I only have one hundred and fifty."

Mahmoud's turn. He shook his head a few times and expressed exasperation, telling me, in effect, that I was not playing by the rules. I was not bargaining properly. I disagreed.

"Of course I am discussing properly. I am bargaining," I argued. "You gave me a price and I refused it. You gave me another price and I refused that too. All I can pay is one hundred and fifty dollars. That's it. That's my way of bargaining. Do you want me to rent your camels or not?"

He didn't understand my flurry of English words, but he received my message anyway.

"C'est tout?" he asked, his brow furrowed. "C'est tout? Jusqu'a cent cinquante dollar?"

"Oui," I replied, sitting back and waiting. "C'est tout."

"Merde," he swore and looked at me in a new way. "Merde," he said again, a look of astonishment in his eyes. "C'est tout?"

"Oui. C'est tout."

Tinker the Targui

He poured us both another glass of tea. After a noisy sip, he looked at me, questions written all over his face. I smiled at him. He responded in words, without a smile.

"Vous êtes formidable, M'sieu." He shook his head in wonder. "Très formidable."

We drank more tea, served now by the girl who was, he explained, his daughter and not married and who cost him much money to keep.

"She is a beautiful Targuia," I said. That was the wrong thing to say at that moment. He immediately tried to sell her to me.

"You want? Only five hundred dollars," he offered.

"I don't have five hundred dollars," I reminded him.

"One hundred and fifty dollars," he suggested with raised eyebrows and a hint of guile. I felt sure he was joking but played it straight.

"No, I need the money for camels and guides, not for a woman."

Mahmoud grinned at me, perhaps thinking of me as a future son-in-law. "My daughter will be a good woman, like her mother. Give you many sons, Insh'allah."

It was an inspired argument but, as I had not seen any of his wives, and I really did need those camels, I had to forgo the undoubted pleasures of owning a real live Targuia: even though she was advertised as a guaranteed son-producer, and, as I had seen, she was a rather beautiful desert woman.

We parted as friends without cementing either deal. I did not expect him to return but return he did. Mahmoud came to my lodgings the next morning. He had five camels with him, all sturdy beasts, and he had his cousin, Ahmed. They were all ready for the trail. A part of me was hoping that Mahmoud's daughter would join us as well but no such luck. Life is full of disappointments.

"Bonjour, M'sieu Tinker," he greeted me with a smile. "Et voilà, les chameaux." He indicated the camels. They stared at me with bemused expressions as they chewed their cud. I felt they were assessing my ability to survive the rigours ahead.

We had not discussed money but knowing instinctively that he had accepted my price, I took Mahmoud aside and gave him 75 dollars, promising the rest when we reached Abalessa. He accepted the conditions and said he would come back soon. I understood he was going home to leave the money in a safe place. While he was gone I examined the rest of my rented equipment.

Two camels were already loaded with sacks of rice, dried meat, dates, tea and all the little details that make life on a wilderness trail bearable. All five camels carried two fat guerbas, one on each side. The goatskin water bags were full and sweating a little in the rising heat. The water wouldn't taste sweet, but it would be drinkable.

Mahmoud came back and introduced me to a big male camel: my ride for the next few weeks. The stud gurgled deep in its throat and spat at me. He missed.

Watched by Mahmoud and Ahmed, and the other four camels, plus a collection of awed onlookers, I ordered my camel to its knees. He ignored me. I Picked up my staff and showed it to him. He looked away.

I poked it between his hind legs and growled, "Get down now or I'll whack your balls so hard you'll never stop crying."

Having issued the ultimate threat, I aimed a kick at the back of his left front knee. He got the message. With much grumbling and bellowing he lowered his bulk to the sand, never taking his eyes off the staff in my hand. My audience laughed with approval.

Ahmed said something in an Arabic dialect that I understood to mean, "He doesn't like you."

"I don't like him, either," I said in English. "He'll get used to me. We'll get used to each other."

Turning back to the camel, I said, "Okay, Fred. We have a long journey ahead of us. Be a good boy."

Fred, for that became his name for the next few weeks, fluttered long eyelashes at me and grunted. I showed him my staff as a reminder. He growled. Confident that he would obey, I tied my backpack over his rump and he screamed in agony. I told him to shut up.

Standing where he could see me, and my staff, I took off my sandals and tied them to the saddle. Without warning, I stepped on Fred's bent knee with my bare foot and launched myself into the tamzak – saddle. Without hesitation, he got to his feet in the typical rocking and rolling motion of a pissed-off camel. I held on until he settled down. With a wicked looking Tuareg cross in place of a pommel on the saddle, I had no wish to be thrown over it and risk changing my gender on the way.

"Good boy, Fred," I encouraged. "Just remember this." I showed him the staff again. He snarled and tried to bite my toes. I tapped him on his bony head with my stick. For better or for worse, we were now partners. One of us would have to learn to obey – and it would not be me.

Mahmoud and Ahmed mounted up and we rode away from Gao in single file. Mahmoud took the lead, a baggage camel tied to the tail of his mount. I followed. Ahmed brought up the rear leading the other baggage camel. Ambling along at a steady pace we covered an estimated 15 miles that first day. Only four hundred to go, I told myself as I stretched out in my sleeping bag that night.

Fred was in a cantankerous mood again the first morning on the trail to Abalessa. He did not want his hobble removed. He did not want to be loaded. He did not want to do anything he wasn't already doing, which was nothing. He stared longingly back towards Gao. He bit my shoulder when I tried to fit his halter. He spat at me when I forced him to his knees. Instead of completing the manoeuvre, he remained there, like a supplicant at Mass, his rear end in the air, his forelegs bent, his chin almost on the ground.

"You look silly," I told him. "Put your bum down and behave."

Fred cursed me in a gargle of camelarabic. I showed him the stick and turned towards his tail. His bum came down and he settled, couched on the gravel.

"Now he understands you," laughed Ahmed. "That's good."

I folded my multi-hued striped blanket in two. Fred watched. His eyes flickering in annoyance. I draped the blanket carefully over his back and he screamed as though a ton of junk had been forced upon him.

"Shut up," I ordered. He spat in my direction.

I picked up my tamzak and placed it over the blanket. Fred screamed again. I added my takouba. Fred groaned in agony at the additional weight of the sword and scabbard. I reached under his belly and tried to strap the saddle in place. Fred pushed down with all his might making it impossible to use my hands. I withdrew and picked up the stick. Fred became hysterical.

I took his reins in one hand and told him, "Okay. Get up, you big baby."

Moaning and groaning, Fred struggled to his feet and stood there with tears in his eyes. I reached under his belly again and strapped the saddle in place. Fred peed in a gush of steaming yellow water all over my robe and right leg.

"Il est difficile, non?" Mahmoud called with a laugh.

Fred was now dressed for travel. I was beginning to smell like a dockside pissoir in Marseilles. I needed a laundry. I also had to load Fred with his share of water and food. The laundry had to wait. Loading the last items meant getting Fred's belly to ground level again. We discussed it. Fred decided he was happier standing. We argued. He was adamant. I cajoled. He acted as if I didn't exist. I resorted to the threat of the stick on his manhood. His stomach gurgled in ominous displeasure. I stepped aside before the vomit poured out of his mouth. Fred shook his head, licked his lips and winked at me.

Using brute strength, I forced him down. Mahmoud and Ahmed stopped loading the other camels and watched in amusement. Fred did not enjoy being loaded. He acted with outrage as each piece of baggage was tied on. He cursed me. I swore at him. He grabbed the end of my scarf between his teeth and pulled hard, damn near choking me. I threatened him with the stick. He let go, allowing the material to slide over the saliva lining his teeth and gums. It did not improve the odour emanating from my regalia. His expression said, "I think I've won."

He had not!

I finished the loading, perhaps a little rougher on him than I intended. Satisfied nothing could fall off, I picked up the stick and shouted, "Up!" Fred looked at the stick in my hand and raised his rear end. Taking it slowly, he straightened one foreleg at a time and stood, adding a few grumbles for effect. He stamped his spongy footpads a couple of times, and he glared at me, but he had made his statement. Now he would behave, until it was time to unload. Then we would go through the same battle again. And we did, twice each day of our journey.

"Now, you bad-tempered old fart," I tugged at his reins for emphasis. "Let's trek."

I led the way. Fred, docile as a lamb, followed without a murmur.

Yeah! Tinker go walking! Fred go walking too!

And walk we did until we had left our first camp far behind us. Only then did we mount up. Mahmoud led us up into the rocky terrain of the Adrar des Iforas. The swaying of the camels was hypnotic. The hours passed in a collection of dream-like scenes. We stopped at unmarked wells, which Mahmoud located with unerring skill. Working together, we hauled water from the deep in a leather bucket. The water was often brackish; not pleasant, yet bearable when boiled with tea and sugar. It was adequate for our needs.

Mahmoud

We crossed the unmarked border between Mali and Algeria without formalities. Mahmoud pointed to the north-west and said, "Les douanes!" I couldn't see any sign of a customs' post. I shook my head. "Pas necessaire," I answered. I would worry about legalities if or when the need arose. I was dressed in Tuareg robes. My face, hands and feet were dark from the sun. Few would recognize me as a European from a distance.

Ahmed killed a sand viper with his stick. He held it up for me to see. More than half a metre in length, it was the same colour as the sand and rocks. I wondered how he had noticed it. I wondered how many other poisonous snakes I had walked past without seeing them. I wondered how they could survive in that extreme wilderness, even though I knew they preyed on lizards and small mammals – none of which I had yet seen.

The crossing of the rock and pebble strewn Adrar Ilassene was one of the roughest trails I had yet encountered. We rode our camels in the early morning for an hour, and then we walked with them. We stopped for tea and bread at midday, slept for an hour, and then continued, sometimes walking; sometimes riding. And so it went on, day after day between sunrise and sunset. My feet earned callouses on toes and heels through walking in sandals on the rough terrain. Our lips cracked from the heat. Our eyelids stuck together with dried mucous and sand. We spoke rarely, except when sipping our tea, or when I felt it necessary to chastise Fred for being difficult. Conversation took too much effort; too much saliva, which we could not spare.

At night we slept in a triangle formation around a small fire to keep warm after the desert cooled. The camels, all hobbled, were free to do as they wished. Most of them wandered off in the night looking for food. Fred, to his credit, never roamed far. Most mornings when I crawled out of my sleeping bag he was couched close by watching me with his inscrutable eyes. I always broke off a piece of sugar and gave it to him. Not that it improved his mood when being loaded. The other four camels always had to be rounded up in the mornings.

I had tried counting the stars once when camped out as a boy. There were far too many. In the Sahara, with no artificial lights to dim the glow, thousands more were visible. I watched them each night, seeking out familiar constellations. I no longer tried to count the twinkles.

The sun was already low on the horizon over our left shoulders when we came upon Abalessa many days later, with our guerbas now flat and lifeless against the sides of the camels. Our drinking water almost exhausted. Palm trees and the ruins of a once lovely and vibrant community stood in dark silhouettes against the fiery red sky and the glowing orange orb. Mahmoud called a halt and, with a certain amount of excitement, tinged with relief, the three of us couched our camels and dismounted.

Mahmoud lowered his veil and grinned at me through the dust and sand that lined his face. I nodded, removed my own veil and returned the smile from my equally dirty visage. We had made it. We shook hands and made camp for the last time together. My camel, hobbled as usual at night to prevent him wandering away, hopped over and nuzzled my neck. I gave him a lump of sugar.

"Yeah, Freddie, you cranky old bastard. I love you too," I told him as I rubbed his ears. He drooled on my shoulder and nuzzled me again. Our partnership was almost over. In the morning we would bid each other a fond farewell. Fred would soon begin his trek back to Gao with Mahmoud and Ahmed and I would be forgotten.

Yeah! Fred go walking! Indeed, he did.

I stayed because I had a lady to meet in Abalessa, but I did miss that argumentative dromedary.

CHAPTER FOURTEEN

Tin Hinan, Queen of the Tuareg people. Queen of the Sahara. Tin Hinan, a beauty from antiquity. Legend tells of Tin Hinan arriving in the mountainous Ahaggar region of what is now southern Algeria on a snow-white camel and accompanied by her maid – Takamat. That was sometime in the 4th century. It's a fanciful story, almost too romantic, yet it bears elements of truth.

Tin Hinan is revered by the Tuareg as a warrior queen – the beauty from far away, possibly the Atlas Mountains of Morocco, who united the many Tuareg tribes into a single entity based in the mountains and deserts of the Ahaggar region of the Sahara. She sounded like a woman worth knowing. I wanted to meet her.

The departure of Fred and my friends made me sad for a while. I was lonely again. The only cure was action. Time for Tin Hinan. I walked the two miles to her tomb in the cool of an early morning. Planning to stay there all day, I had four bottles of water, some dates and bread.

The walk is not a strenuous one, although the desert is uneven. I took my time to conserve energy as I knew I would need it later for the return hike. Wearing my hiking boots instead of sandals, and with my trusty wooden staff in hand to steady me against the possibility of a fall, I covered the distance in less than thirty minutes.

From a distance the tomb of Tin Hinan looks like nothing more than an irregular hill against the backdrop of the jagged Ahaggar Mountains. Closer, details become visible. Details that show the work of mankind. Up close it becomes obvious that the burial mound is enclosed in a man-made wall of desert stone. It resembles a fort.

Archaeologists have written that the design and construction of the burial mound is not typical of the Sahara. Not even African, in fact. They believe it is possibly Roman in origin. Among the artefacts excavated from the ruin were Roman coins, lamps and shards of pottery. Those important clues pointed to the 2nd and 4th centuries A.D. Takamat and her fellow Tuareg must have buried their beloved queen deep inside an existing fort. Tin Hinan's skeletal remains and the goods buried with her, such as bracelets, necklaces, and coins, have since been exhumed and sent to the Bardo Museum in Algiers.

I was happy to spend the day at the tomb site, even though Tin Hinan was no longer there. I felt her presence anyway. Many years later, wondering about Takamat, I tried to find out what happened to her without success. Considering her importance to the Tin Hinan story, I found that lack of information extremely sad.

The village of Abalessa exists, not because of Tin Hinan's tomb, but for the fertility of the region. Abalessa has in excess of 300 healthy date palms irrigated by foggaras, or man-made underground canals. They are said to be the longest and deepest in the Ahaggar. I climbed down a wooden ladder into a foggara with a local and estimated it to be a little over 30 feet deep. It was cool with naturally chilled water flowing in a narrow river. It was also claustrophobic, and it was populated by lots of spiders. I didn't stay long.

Soon after arriving in Abalessa I advised the head man that I had, *je m'excuse*, somehow missed the border post when coming from Mali. Would he be so kind as to stamp my passport and officially welcome me to Algeria? He wasn't thrilled to receive my request, perhaps the first of that nature imposed on him. He gave an emphatic, "No!" I pleaded with him, resorting to every sad facial expression I could think of – short of tears – until he agreed to decorate my passport with his jealously guarded official Abalessa stamp.

"You will have to go to Tamanrasset and speak to the gendarmes there for an entry permit and for permission to cross the desert," he warned.

"Would the gendarmes at In Salah be able to process me?"

"Hmm. Je ne sais pas. Peut être." I do not know. Perhaps.

In Salah was halfway across Algeria. It was the northern end of the Trans-Saharan piste. Where could they send me if they chose to bar me from entry? Not south, for that would entail supplying me with a permit anyway. I liked the odds in my favour. In Salah it had to be, if I could get there. Help was on the way.

TINKER GO WALKING!

CHAPTER FIFTEEN

Two Frenchmen in an old Citroen Deux Cheveaux arrived at Abalessa, their lightweight car rattling like a bucket of nuts and bolts on a vibrating table. They were both at least five or six years older than I. They were in trouble and they were desperate. I saw a ride across the Sahara to the northern fringe.

"Voulez-vous desire assistance?" Do you need help? I offered. My accented French betrayed my origins.

"Vous ets Anglais? Americain?" one asked.

"Anglais, et Canadien. La même," I answered.

"You are un mechanique?" one grasped at the straw, hesitant in a mix of English and French.

"Yes, I am. If you have a few tools, I can fix anything. I'm a tinker."

While they explored the ruins on foot, I spent three days with their car, most of that time with the simple engine in pieces on a square of tarpaulin. Its sensible simplicity reminded me of the German Kubelwagon I had repaired at Aswan.

I cleaned all the parts, washing them in a bath of motor oil I found thoughtfully stowed in the car. When I put it all back together again and put the key in the ignition, the Frenchmen held their collective breath. So did I, but I did not show it.

I tapped the gas pedal a couple of times, pulled out the choke, turned the key. The engine coughed and wheezed. I turned the key off, pushed in the choke, turned the key on again. The engine coughed once more… it sneezed, and then it began to purr. I wiped my oily hands on an even oilier rag and grinned at the owners.

"Ca marche," I said. It goes. The Frenchmen, Claude and Raymond, applauded. They had come from Algiers and made the side trip to Abalessa en route to Tamanrasset. I was happy they had done so. Now they were in a hurry to move on.

We stayed one more day in Abalessa while they explored the oasis and then we started west for the main Trans-Sahara piste. My staff projected out of the open roof like an elongated unicorn horn. I soon realized why the car had broken down. The Frenchmen had no idea how to drive in a desert. I offered to take over when we reached a tiny historical village with the delightful name of Tit.

Tit is one of those remote places that history has all but ignored. Some French people know about it. Almost all Tuareg know it well. Modern Sahara travellers should know of it. Tit has a violent history.

In May 1902 a French military expedition defeated a Tuareg force at Tit. The Tuareg, who had initiated the battle, were armed with swords and lances and fought on camelback. The French were armed with up-to-date weaponry. The Tuareg force suffered crippling losses while the French lost only three men and a few wounded.

Legend says that the Tuareg always turn their heads away in shame when passing the site of their worst defeat. In homage to the Tuareg, thinking of my friends Mahmoud and Ahmed, I turned my head away too.

Tamanrasset is half a day's drive up into the Ahaggar Mountains. It nestles in a bowl surrounded by sharp volcanic peaks. Tamanrasset in the middle of the Sahara, a small town populated by Tuareg, most from the two dominant tribes – the Dag Rali and the Adjoun Téhélé.

A Targui, robed, veiled, wearing a sword at his waist and sandals on his feet walked down the middle of the main street in Tamanrassset. That in itself would not have been cause for comment, there are many such men in and around Tamanrasset. This Targui was notable because he was pushing a garden wheelbarrow in which stood a complete white porcelain toilet bowl and cistern. What is he going to do with that? I wondered. Does he even know what it is for? There was no way of telling. The images swirling in my mind were as comic as they were gross.

The Targui strode past me without a glance in my direction and walked off into the desert – pushing his loaded wheelbarrow.

The two Frenchmen had driven across the Sahara primarily to visit the hermitage of Père de Foucauld at Assakrem. Their Abalessa visit had been a fortunate side trip for me. Visiting Tamanrasset and driving around the Ahaggar Mountains was a wonderful bonus.

I had never heard of Père de Foucauld until Raymond started talking about him. The more I learned the more I wanted to visit the place where the French missionary had lived, prayed, and where he died.

Charles de Foucauld was born in Strasbourg in 1858 into a Christian family. He became a soldier at 20-years-old and served in North Africa, where he became enamoured of the land and the people. After his military service he settled in Algeria for over a year to prepare himself for an extended journey in a forbidden land. At that point in his story my interest in de Foucauld went from mild curiosity to a desperate longing to know more.

Morocco was closed to Europeans. Any who dared enter the forbidden land had to do so in disguise. Charles de Foucauld roamed Morocco for close to one year, learning about the country, the people, and their religions. He returned to Algeria in rags and close to starvation in 1886. During the next few years de Foucauld travelled in the Holy Land, he spent seven years as a Trappist monk and a further three years as a hermit in Nazareth. On returning to France in 1890 he entered a convent and was ordained a priest the next year. Père Charles de Foucauld was en route to his destiny.

He requested a posting to Beni Abbes in Algeria, a small community nestled among the great sand dunes, and stayed there for a few years. In 1904 he moved south having volunteered to live among the Tuareg and teach them the importance of Christianity. He spread the word for ten years from his Hermitage at Assakrem yet, by his own admission, failed to convert any of the Tuareg to Christianity. Père Charles de Foucauld was murdered by Islamic religious extremists on December 1, 1916.

Leaving Tamanrasset just before daylight, we drove into the stark beauty of the Ahaggar. The track was littered with stones, large and small, requiring careful, slow driving. The approach to Assakrem is a series of narrow switchbacks as the track climbs to around 5,000 feet above sea level. We parked the car and clambered the final few hundred feet up a rocky path to the Hermitage.

Inside de Foucauld's hideaway, built by hand from local stone, is the altar where he prayed. A crucifix hangs on a rough-hewn stone wall lit by a shaft of sunlight. There was something intensely spiritual about the tiny room. The three of us knelt as one and bowed our heads. I don't know if Claude and Raymond prayed that day, but I certainly did.

We drove back through the rocks and gravel, the sheer mountains and the boulders, on a different route that allowed us to complete a circle around the area. Apart from occasional comments about the difficult road, we drove in silence. My mind was on the solitary life of the ill-fated French missionary.

Claude and Raymond already had a permit to take their car north across the Sahara to In Salah. We decided that permission should encompass all three of us as well and set off for In Salah after topping up our fuel tanks and water.

Driving north, I stayed off the corrugated gravel and sand trap that was the Trans-Sahara piste. Instead, I ran the underpowered car parallel to the 'road' a hundred yards or so to the west. We made good time, camping out one night about halfway. Under my care, the old 2CV purred happily all the way across the desert to the raspberry red town of In Salah. All travellers crossing the Sahara from Tamanrasset to the north are obliged to report to the Gendarmerie at In Salah. The officer in charge checked my passport, found the Abalessa stamp and added his own beside it. Well over a month after I crossed the border from Mali, I was officially welcomed to Algeria.

Outside the Gendarmerie Claude called our attention to a rare sight. Coming towards us was a camel pulling a large four-wheel flatbed wagon the size of a car. Tied on top of the wagon was a six-cylinder truck engine. A thumbs up for desert ingenuity and a thumbs down for the wonders of modern machinery.

I parted from Claude and Raymond the next day at El Golea. By then the road was paved and would be so all the way to Algiers. They didn't need me anymore. Claude shook my hand and thanked me for my help. Raymond hugged me and kissed me on both cheeks.

"Merci, Tinker," he said, pronouncing my name as 'Tinkair.' "Merci."

"Bon chance," I called as they drove away.

It occurred to me that day that I had always shied away from physical contact with another man, even a close friend, other than to shake a hand. Raymond's spontaneous gesture showed me there was nothing wrong in a healthy hug. I have never been shy about hugging my friends of both sexes since that day.

For the next week I roamed the great dunes of the western sand sea – the Grande Erg Occidentale – from El Golea. Some nights I camped out alone under the stars. On other nights I shared a fire with nomads. I was at peace.

The realization that my Algerian visa was due to expire in a few days spurred me to hitch a ride north to Ghardaia, a lovely 11th century town in shades of red and white climbing in terraces through narrow streets to the proud minaret of the mosque. I wish I could have stayed. I vowed to return. In a hurry to leave the country, I crossed into Tunisia and caught a ride to the Libyan border beside the Mediterranean Sea.

There is a ruin I wanted very much to visit in Libya called Leptis Magna. Said to be one of the most beautiful cities of the Roman Empire, it sprawls along the coast overlooking the Mediterranean east of Khoms. I had hoped to obtain a visa at the border, as is possible in so many parts of northern Africa. I was disappointed. The Libyan border guards at Ras Jedir refused me entry, insisting I go to their embassy in Tunis and apply for a visa there. A German traveller, with the correct authorization in his passport, told me he had waited three weeks in Tunis before the Libyan Embassy granted him a visa valid for seven days. That was too long to wait for too little time to explore. Leptis Magna went on my list of places I would return to one day.

I hitch-hiked up to Tunis where, on principal, I enquired of the Libyan Embassy how long it would take to obtain a tourist visa for one month. The answer was as predicted: "We only issue visas for one week, and there is a long waiting list."

Two more important archaeological sites remained on my mental list for northern Algeria. I renewed my visa at that embassy in one day and made my way west on winding mountain roads for Timgad. Rides were plentiful. The drivers were courteous and, without exception, maniacs behind their steering wheels. I have never prayed so often or so hard as I did on the roads to Timgad.

A Berber village named Thamagudi once occupied the site now obscured by Timgad, the foundations for which were laid in 101 A.D. during Trajan's tenure over the Roman Empire. Walking the geometric pattern of streets, just as tedious as the grid system used in North American towns and cities, I tried to imagine the city filled with Roman soldiers, Berber warriors, housewives, slaves, children, seamstresses, bakers, tailors and, yes, of course, tinkers – the blacksmiths of the time. The images refused to focus. In most cities of Roman, Greek, Persian origin, different ethnic groups gathered to live and work in clearly defined areas. Artisans did the same. I could see no evidence of any such congregations at Timgad. There was no hint of spices in the air. No brash sounds of copper or iron being forged into useable shapes. No cooking smells from open-air braziers. No reminders of the bustle of humanity. Devoid of human life, the rigid geometric layout gave the ancient settlement a sterile appearance. Unlike Ephesus, or Persepolis, to name but two attractive ancient cities, there was no warmth in Timgad.

Only the Forum and the Theatre offered any break from the mathematical discipline of the streets. And yet, I accepted, those two important features of Roman life were equally geometric. The Forum being a large rectangle; the Theatre a semi-circle of rising terraces facing a rectangular stage. It occurred to me that Timgad would be a fascinating place for a mathematics teacher to take students for a field trip – geometry on graphic display.

Timgad interested me and I was happy to have taken the opportunity of walking its almost 2000-year-old streets, but it left me cold. Even the remains of the Byzantine fortress south of town failed to impress. Perhaps it wasn't Timgad. Perhaps it was me.

My plan was to visit Djemila on my way to Algiers. The sterility of Timgad almost made me change my mind. Perhaps I had seen too much. Perhaps I had savoured too many exotic experiences in too short a time. My brain felt as though it were overflowing with intellectual sensations. It was time to go home for …whatever home entailed.

Djemila was not far from a main road. If I could catch a ride to the site without waiting more than an hour or two, I would do so. Fate looked after me that day. I left a truck at El Eulma and was picked up in an old car minutes later. That ride took me within an estimated five miles of Djemila. I walked from there.

The mountain air felt good after so many months in the desert. My hiking boots, now free of irritating grains of sand, accepted the paved road gratefully. I stepped out at with an easy stride, my trusty wooden staff in one hand. The time and distance passed in pleasurable paces.

Djemila, or Cuicul, as it was known to the Romans, is a fine example of a Roman town planning attached to a fertile hillside. I looked down on the remains clinging to the steep land from the advantage of a south-western prominence and appreciated the strategic positioning – and the panoramic view.

The town was built at the end of the 1st century A.D. on the side of a rocky outcrop overlooking two wadis, or river valleys. It, in turn, is watched over by three heights. Djemila was a military garrison, manned, it appears, by veterans. History has recorded the peaceful co-existence of Roman soldiers, Berber tribespeople, farmers and traders. Laid out almost on a north to south axis, Djemila has two main streets that converge at the north end of town.

Road access could only be from the northeast as the south of town butts against a cliff – an almost perfect defensive position. I sat on the front row of the second tier of the theatre for a long time, with my back to the dark schist of the rock wall, watching the changing light as shadows played over the adjacent buildings of ancient limestone.

Ahead of me the stage was bare. There were no actors waiting in the wings. No audience clamouring for seats. Despite the lack of theatrical excitement, the theatre has defied the centuries with only an occasional green weed poking its head between the flat stones of the stage to record the passing of time. To my right the considerable remains of a triumphal arch stood as a testament to the talents of its architects and the skill of stonemasons. Like the theatre's stage, and the streets of Djemila, the arch had endured winter storms of snow and ice. It has withstood the excessive heat of the North African summer, and it stands today as a vivid reminder of the incredible scope of the Roman Empire.

I departed Djemila feeling much more at ease than I had after my stay at Timgad. My feet and my staff took me downhill for most of one day until a friendly driver gave me a ride to Sétif. After a night on a park bench of cold stone, I took a bus all the way to Algiers – my departure point from Africa. There I debated whether to visit Tin Hinan's remains at the museum. Somehow the idea seemed lacking in respect. I left the long dead lady in peace.

In the past year I had roamed through Egypt, Sudan, Tchad, Cameroun, Nigeria, Niger, Upper Volta, Mali and Algeria, plus I had raced through northern Tunisia. It was time to say goodbye to Africa. When I left Algeria I was mentally exhausted and ready (I thought) for a normal life. I was wrong, of course.

CHAPTER SIXTEEN

From Algiers I crossed the Mediterranean to Marseilles: a day and a night on a ferry that needed a few months of spring-cleaning and an extended visit from a team of pest control experts. A long haul by a succession of private cars and busy trucks then took me the length of France to the Channel coast. There, at a rainy and wind-swept Le Havre, after a few days of waiting, I boarded the small Dutch ocean liner, the S.S. *Ryndam*, bound for Quebec City and Montreal.

The ocean voyage was a philosophical panacea for my mind. The rolling Atlantic waves, the wind that played with the ship from all points of the compass, the sensation of being wonderfully lost – as among the great dunes of the Sahara sand seas, became a buffer between what was past and what was now, and prepared me for what was to come.

We approached the Grand Banks off the south-east coast of Newfoundland in thick fog, listening every few seconds to the mournful howl of the ship's deep whistle. Most passengers retreated to the comfort of the lounges to read, play cards, drink, or to their cabins, to avoid the damp. I stayed on the promenade deck and was rewarded with the sight of a whale passing close by. Subsequent research told me it was a North Atlantic right whale. I saw nothing else, other than the grey miasma shrouding the sea.

Although I was going home and should have felt something akin to excitement, the dominant thoughts in my mind were of my travels. Had I found my Lilliput? Had I found my Brobdingnag? Well, I had met little people on my peregrinations: some with big minds; some with less. I had met big people: some with little minds. Some with much more. I was no Lemuel Gulliver but, I think I already knew, the destination was, for me, not the most important aspect of travel. Getting there was the challenge. The destination was just a small reward.

Daybreak dawned in the Gulf of St. Lawrence. We steamed out of the fog and into early morning sunshine. Passing the Gaspé peninsula on our left we cruised into the majestic St. Lawrence River. Soon the banks of farmlands gave way to tidy villages with white-washed churches topped by slender red steeples. Quebec City, where I first set foot in Canada, came in sight, heralded by the Chateau Frontenac towering over the charming old city. I went ashore in the early evening, climbing the scores of wooden steps to reach the Plains of Abraham – an 18th century battlefield where British and French forces once clashed. I sat there in solitude for an hour, enjoying the lights over the river.

Another night on the ship cruising up the St. Lawrence and we docked in Montreal. After the organized chaos of disembarkation, a railway train – the final leg of my odyssey – carried me overnight to the cultural void of Toronto in late 1960.

Tinker had gone walking. Tinker had come home. For how long? I wondered, thinking of a distant echo from long before which reached me as I tried to sleep one night on the ship on my voyage back to reality…

"Where's Tinker?"
"He's gone."
"Gone where?"
"Oh, gone walking, I suppose. Gone to see the world. Timbuktu, or someplace like that, he said."
"Will he come back again?"
"Yes, I expect so, one day."

I did come back to find nothing had changed, except me. I had changed forever. History and the world had seen to that. Nothing of my past could ever be the same. Tinker had gone walking. A different Tinker had returned. He was older; wiser, but a tinker still.

They were all pleased to see me at first – everyone in our family, that is. It didn't last. I had changed too much. I was two years older. They were two years older, too, other than that, they had changed not at all. Mother could still pick a fight in an empty room. The old man still didn't have any answers. Within a few weeks I knew I had to move on. Places to go. People to see. I had a life to live.

Yeah! Tinker go walking!

TINKER GO WALKING!

EPILOGUE

In those days Tinker could not settle anywhere for long. The open road called too often. It called too loudly. Somewhere beyond the seductive hills a new adventure awaited, beckoning him onwards. He had no choice but to follow.

Some said he joined a rock 'n' roll band and travelled the world as a singer. Others, more astute perhaps, claimed he went to work on the oil rigs in Kuwait. There is evidence that for many years he owned and operated an expedition company working in the Sahara and the deserts of the Middle East. A rumour, not so far from the truth, said he was last seen on foot leading a Bactrian camel into the blistering heat of the Dasht-I-Margo of southern Afghanistan. Another said, no, he was studying Royal Bengal tigers in the Sundarbans jungle of Bangladesh. Someone claimed he had drowned in the Arctic Ocean. Some were right; some were wrong. Arguably the best indicator of Tinker's later life can be found in a journal in which he states:

> My long journey of discovery through fabled desert lands in my youth became a guide for my future. I believe it defined my destiny. It helped me become a passionate writer. Many more adventures followed, in deserts, on the oceans, in jungles, on mountains and on polar ice. A host of, mostly

self-inflicted, near death experiences shadowed the middle years of my life. They were years when I fell in love. And years when I fell out of love, and years when I fell in love again. Despite the rumours, and despite the conjecture, I survived them all. If you believe that I am indeed both Tinker Taylor and Anthony Dalton, then you should know that twenty-one of those stories have been told in *Adventures with Camera and Pen*. Two of them became books of their own. They are: *ALONE Against Arctic Seas* and *River Rough, River Smooth*. A few others have formed the basis of my novels and short stories.

Now, many decades after that initial long walk into the wide world in Scotland as a tiny boy. Long after the gypsy tinker and his exotic lady have faded into the recesses of my memory. Long after Molly (and her passion) has grown old. Long after Brandy and all the others who followed have lived out their lives in their own worlds, I have found a new gypsy lady. She wears a flowing skirt, trimmed with black lace and decorated with a ruby red and emerald green bohemian pattern. She matches the skirt with a ruby red blouse and a short black jacket with long sleeves. Her hair is tied up in a bun wrapped in a ruby red and emerald green silk scarf. She likes ruby red, and emerald green, and cerulean blue. She adores strong colours. So do I, especially when she wears them.

I still have my trusty walking staff and I still like to walk, preferably with a dog for company. I continue to feel a thrill at the sight of a narrow winding country road, a signpost, a rolling hill, a forest of new growth reaching for the sun, a clump of smiling wild daffodils, or when I hear a clear stream giggling over rocks. At these times, with the image of a lovely gypsy-like lady in my mind, my heart sings out with the joy of our world.

"Yeah! Tinker go walking!"

The End,
and a new beginning.

Tinker the elder

TINKER GO WALKING!

ABOUT THE AUTHOR

Anthony Dalton is a Fellow of the Royal Geographical Society and a Fellow of The Royal Canadian Geographical Society. A life-long adventurer, he is the award-winning author of 16 non-fiction books – most about the sea or about exploration – six novels and a series of short stories. A past president of the Canadian Authors Association and an accomplished public speaker, he is an historian and a former expedition leader. He lives in the Southern Gulf Islands of British Columbia.

TINKER GO WALKING!

More books from Anthony Dalton

Fiction

Sarah's Mountain

Albert Ross is Lonely,

A Diamond Quintet

The Sixth Man

The Mathematician's Journey

Relentless Pursuit

Non-fiction

Henry Hudson

Sir John Franklin

Polar Bears

Fire Canoes

The Fur-Trade Fleet

A Long, Dangerous Coastline

Graveyard of the Pacific

Arctic Naturalist

River Rough, River Smooth

Adventures with Camera and Pen

Alone Against the Arctic

Baychimo, Arctic Ghost Ship

J/Boats Sailing to Success

Wayward Sailor

Short stories

Found!

The Twelfth of Never

When the Bluebird Sings

A Voice in the Wind

Alexa Stole My Wife

Jungle Predator

Jensen's Lion

Screenplays

Albert Ross is Lonely

Whiplash

Infinity is Forever

The Mathematician's Journey (TV pilot)

TINKER GO WALKING!

ADVENTURES WITH CAMERA AND PEN

ANTHONY DALTON

ADVENTURES WITH CAMERA AND PEN

Anthony Dalton's adventurous photo-journalism assignments and expeditions have taken him from the Arctic to Afghanistan, from the Outback of Australia to the Falkland Islands, from the Namibian desert to the Sahara, and from Bangladesh to the Empty Quarter of Saudi Arabia and Oman. His journeys have taken him into the mountains, jungles, across deserts, and on stormy seas.

Join him as he sails as a crewmember on a massive Russian tall-ship, tracks Royal Bengal tigers through the Sundarbans jungle, gets mixed up with exotic wildlife in southern Africa and the Falkland Islands. Through it all, whether fishing in the Arctic or climbing in the Bugaboo Mountains, Dalton's irrepressible sense of humour shines through in his writing. A great read for outdoor adventurers of all kinds, and for armchair adventurers everywhere.

ALONE

AGAINST THE ARCTIC

Anthony Dalton

ALONE Against the Arctic

In the summer of 1984, Anthony Dalton embarked on a near-fatal voyage in a small boat along the wild northwest coast of Alaska, attempting a solo transit of the Northwest Passage. His se quest ran parallel to an arduous relief expedition undertaken in 1897-98, when the officers of the U.S. Revenue cutter *Bear* set out to reach eight whaling ships that were stranded in thick ice, their crews on the verge of starvation. Both journeys are depicted in this captivating adventure tale, and Dalton's gripping description of his encounter with an icy hell explores the irresistible lure of risk and challenge that continues to draw adventurers to the Arctic, a place like no other.

RIVER ROUGH, RIVER SMOOTH

Adventures on Manitoba's Historic Hayes River

Anthony Dalton

RIVER ROUGH, RIVER SMOOTH

Manitoba's Hayes River runs over 600 kilometres from near Norway House to Hudson Bay. On its rush to the sea, the Hayes races over 45 rapids and waterfalls as it drops from the Precambrian Shield to the Hudson Bay Lowlands. This great waterway, the largest naturally flowing river in Manitoba, served as the highway for settlers bound for the Red River colony, ferrying their worldly goods in York boats and canoes, struggling against the mighty currents.

Traditionally used for transport and hunting by the indigenous Cree, the Hayes became a major fur-trade route in the 17th to 19th centuries.

This is the account of the author's invitational journey on the Hayes from Norway House to Oxford House by traditional York boat with a Cree crew, and later, from Oxford House to York Factory by canoe in the company of other intrepid canoeists – modern-day voyageurs reliving the past.

ALBERT ROSS IS LONELY

Anthony Dalton

Praise for *ALBERT ROSS IS LONELY*

Albert Ross Is Lonely by Anthony Dalton is an eloquent, evocative tale, simultaneously grounded in a tender and realistic love story and soaring with fine flights of allegory. Among the gannets, the gulls, and the guillemot colony up on the cliff resides the lone black-browed albatross (appropriately named Albert Ross) in the northern hemisphere. The journeys of Albert Ross are chronicled and photographed by Tripp, who came to the Scottish highland coast to study the majestic bird and to live out the last of days (his heart condition rendering his 48-year-old body in a weakened state). There Tripp encounters the lively and curious ornithologist Amanda. The exchanges between the cranky yet clever Tripp and quick-witted, disarming Amanda serve as one of the novels many pleasures. Dalton has a fine ear for dialogue and the delicate building of affection between the two is subtly developed.

As a young sea captain looks at the albatross, he starts "thinking about how lonely it must be – the only one of its kind as far as he knew for thousands of miles." That description could be applied to both Tripp and Amanda: each of these free spirits has a deep appreciation of nature and a great willingness to sacrifice the comforts most of us require to explore its breath and depth. Albert, Tripp, and Amanda are wonderful company for the reader. The rhythms of their individual and shared journeys ride the thermals as far as nature's forces allow.

I highly recommend *Albert Ross Is Lonely*. The last scenes manage to be concurrently inevitable and surprising – true to nature and true to these two individuals who for a time inhabit the rugged Scottish coast. Dalton is a smart and sensitive writer who has given the reader deeply invested characters and has written with the same integrity and commitment as those explorers who inhabit his fictional world.

Michael Hartnett

Author of – *Fools in the Magic Kingdom*

The Mathematician's Journey

ANTHONY DALTON

Praise for *THE MATHEMATICIAN'S JOURNEY*

Anthony Dalton is himself no stranger to the Arctic or sailing. He brings that knowledge to bear in his meticulously researched historical novel to take the reader on a voyage of discovery from the peaceful cloisters of Oxford of King James I's England to the Atlantic wastes and the frigid waters of what will come to be known as Hudson's Bay thence to a native village and finally, after many years, back to upper class London. Peopled with three-dimensional characters set against vividly described backgrounds the plot will keep you turning the pages and at book's end hoping the hero's adventures will continue in another work.

Patrick Taylor

New York Times best-selling author of the *Irish Country Doctor* series.

TINKER GO WALKING!

For more information please scan and
visit Anthony Dalton's website.

TINKER GO WALKING!

Made in the USA
Middletown, DE
23 August 2019